A KNOWLEDGE ODYSSEY

Why All Atoms are Similar

Why the Universe Expansion is Logarithmic

Through the Portals of Time we Behold Knowledge

Also by David Halsey

Halsey Genealogy Since 1395

Uphill Both Ways-Barefooted

It's All About Time

Goin' Back in Time For a Visit With Ancestors

A KNOWLEDGE ODYSSEY

Through the portals of time we behold knowledge

Slide rule vs Computers

Earth's Regional Climates

Why all Atoms are Similar

The Universe expansion is Logarithmic

First Printing: 2015

Printer: Lulu.com

ISBN   978-1-329-64065-8

Publisher: Colonel David H. Halsey P.E.

US Army Corps of Engineers (ret USAR)

www.g.dave.colorado@gmail.com

Contents

ACKNOWLEDGEMENTS
Anything you can imagine you can be!

The very essence of this book was brought about by everyone that entered "my world". A single person does not exist on an island in a world of humanity. Life's successes and failures involve an infinite collection of people that meander throughout an individual's life. We are never alone.

This dedication was taken, in part, from my acceptance speech upon being inducted into the Mullens High School (West Virginia) Hall of Fame on August 15, 1987.

Every person that I have ever met deserves some credit for my ability to write this book. Some obviously deserve a bigger part than others. Much credit goes to my teachers: Mrs. L. T. Lail (first through third grades in a one room school that my father attended in Pierpoint, WV); Sis Murphy (fourth grade at Maben, WV); Maude Hypes (fifth); Hershel Shumate (sixth); Jannette Houck (seventh and eight); and Ted Clay (principal Maben Grade); and from Mullens High; Mrs. Staats, Mr. Toler, Coaches White and D'antoni, Lee Hodson (physics), Madge Kaman (chemistry), Bill Murphy (woodworking shop), and Loutisa Morgan.

Loutisa has knew me longer than anyone because I begin this adventure on April Fool day, 1935, in her house on a hill in Pierpoint, Wyoming County West Virginia, across from the Primitive Baptist Church. She taught my father in 1920s and me in high school in the early 1950s. She married my mother's first cousin, he died early then she married my dad's first cousin. She retired from teaching at Mullens High School and was the inspiration to me becoming a member of the school hall of fame.

If "Uncle" Charlie McGraw, the one-legged shoe cobbler janitor at Maben Grade School, had not taken the time to show a 13-year-old kid how the steam furnace worked (he always called me "King" David). I was a naïve kid who was awed by the school's indoor plumbing (our houses were not plumbed until I was away in college) and steam heat.

If Mr. Clay, the principal, coach and roll model, had not taken the time to show me how and why he signed a check a certain way so the bank would cash it, or, if he had never allowed me to be the teacher for the first and second grades when their teacher was sick or snowbound; or, if Herbert Morgan (Dad's first cousin) had never allowed me to do the pricing calculations that were required for grocery stores, following W.W. II, for commodities that he received for his Pierpoint store; my leadership and accountability traits may never have been honed to allow future responsibilities. They each trusted me in every way.

I've had the good fortune of maturing during a time of unprecedented technological advances. From the time I entered Mullens High in 1949, became a professional engineer and started to make a contribution, until 1975, a span of a mere 25 years, more technology was created than in all time before. I was at the very leading edge of the hydrologic mathematical modeling world. With large computers and techniques using finite element techniques, we were solving both sides of the energy equation, that defined moving water as mass and momentum, which had not collectively been done before with the technique and methodology that we and others developed. With these tools we could forecast how the Ohio and Mississippi Rivers would react in real time flow forecasting, more accurate and reliable and for a much longer future period than before.

My number one mentor and the most intelligent person that I have ever encountered made it to the eight-grade at the one-room school that I later attended. He was Ralph Halsey, my father 1909-1983. His armor had no flaws. Now at age 80 I see my father every time I look into a mirror. *I have felt his hand upon my shoulder all my life. He smiles!*

When you "boil" down the essence of my basic wherewithal, you'll find that I'm just a coal miner's son doing the best that I can with engineering, physics and military foundation and an imagination as a catalyst.

PREFACE

Knowledge is: (1) the fact or condition of knowing something with familiarity gained through experience, education or association; (2) acquaintance with or understanding of science, art or technique; (3) The circumstance or condition of apprehending truth or fact through reasoning; (4) the sum of what is known; (5) the body of truth, information and principles acquired by humankind. Webster

I entered Marshall College in 1953, received a commission from the Army ROTC program in 1958, and graduated from Marshall University with a Bachelor in Engineering Science (BES) in 1961. I studied engineering with a slide rule but never used one professionally. I walked into a career that was driven by computers, thus continued learning in a technical changing environment was essential for success. This book highlights some of my favorite subjects that I care to share with the reader who is familiar with science, as well as to the engineer and physicist.   Accordingly, to keep the text uncluttered, the sources for quotations from other factual information are foot noted where applicable.

As I study astrophysics in more than a score years in retirement I find that classic formula, equations and theories are beginning to be suspect, and that future scientists need to discover new ways to define the microcosm and macrocosm.   As time progresses in the Universe the classic laws appear to change also.

The current state-of-art in science cannot define why, what or how 'life' exists. Our concepts of life, from birth to death, are constrained by having only observed a single instance of evolutionary life. Our brains have not evolved to the point where original thought can occur. The brain is preprogrammed (learning) and relies upon synergy techniques to accumulate knowledge. To be creative or an inventor are misnomers since we simply discover Nature's secrets, and they are absolute! Marshall engineering students will encounter and define these mysteries. Reality and experience has shown us that "no one" has ever predicted the future. The predictors have neither the imagination nor forethought, just hindsight!

Many of the philosophies that shape our societal characteristics can be traced back 2600 years to the Greeks. That was the time when individuals started questioning whether or not the "gods" were responsible for controlling nature's events. They had a "god" for every natural event that occurred. Many societies have risen and fallen since that time; however, the more each of those societal perceptions altered human thoughts about natural reality, the more the interpretation of societal behavior as a whole has remained the same or similar. From this knowledge, one could conclude that the behavioral attitudes of societies as a whole are "going forward to the past".

The culprit could be that the genetic capacity or make up of the preprogrammed Homo sapien brain which has not appeared to have evolved for millennia. Maybe Nature has concluded that there is no need for the brain to evolve beyond its original form since there will ever be a time when Homo sapien will tax the basic survival and propagation instincts that control the human entity.

Since the beginning of the Homo sapien species, there has been an influx of knowledge about natural phenomena. Some individuals know more about science that defines Nature today than they knew yesterday but does society as a whole know more or care to know more about Nature today than yesterday?

So it can be written that in 2600 years our society has merely traded "Greek gods" for 21st century "causes" or "politically correct" attitudes.

My objective is to create a minute crack in the seam of understanding how Nature really works.
One final caveat: the technical topics are presented as simply as possible. Science is not based upon a belief system, but upon knowledge that evolves over time. What is germane is the data collected, analysis of that data and the credentials of the person(s) making the analysis, and how the results are presented. These findings can be added to your knowledge "banks". In every case the "belief" is never "yes or no" but "here is the results from so and so…

These presentations are gleaned from an 80 year knowledge odyssey that continues. Knowledge is the greatest treasure that we can accumulate. Engineering, physics and military knowledge are my foundations that have expanded into similar, but not same, outcomes.

Following an introduction of how technology advances and thus affects one's career are three chapters that contains real time analysis of data to solve hydrologic/water resource operational practical problems that I want to present to the reader.

Next we begin with a discussion about what we know about the flow of time. It's time that theoretically makes everything possible by not causing all events to happen at once. When we looking at the Periodic Table we see atoms (elements) arranged as to "how" they interact, not "why" they do. The discussion in the chapter on atoms centers upon the why.

Continuing with time we know that every entity has its own local time. 99.99% of an atoms matter is contained in its nucleus. Thus the only difference that any atoms have from the first atom, hydrogen, form after the Big Bang, is the local time within its nucleus. This makes the local time, for the hydrogen nucleus, approach "zero as a limit". Thus all atoms after hydrogen have their own local time which is faster by some minute amount. My theory is that time is the catalyst that determines whether an atom is iron, gold, etc. or some other atom by altering the ratio of matter's states, e.g., mass vs. energy. The minuscule increases in local time converts energy into mass. The lineage and mechanism for time changes in atoms are discussed.

Following the time chapter is a discussion of how the Universe expands logarithmically. We have no idea how big, old or what shape defines the Universe except it's large and old. We have only observed a minuscule portion of the Universe. The more we study the Universe the more we believe that its laws are complicated, yet simple, and are changing with time. Observing the Universe is looking back in time from within the Universe that is increasing in time as it expands. What complications does this pose? Time also turns Universal matter into mass from the same energy (dark matter?) that causes it to expand thus the Universe is growing with time, or is it? All of these thoughts are discussed.

Physics is about everything we see and do, from fixing a flat tire to this illusion we call reality. Over time I have found that an engineering/science/military mind fits my persona perfectly. Yet most people come to the printed page knowing about life in the terms of tragedy, triumph, love and loss but not about theories describing viruses or galactic space. Hopefully, these stories can bridge a portion of this gap.

Outside the boundaries of their respective domains, atoms, Universal expansion, carbon dioxide, Earth's regional climates and water management are defined in detailed description that a novice scientist should be able to decipher.

The Epilogue gives my "take" on the whole "shebang".

Some[1] believe the basic structure of the Universe appears to favor the creation of complexity: e.g.-disordered energy states produce atoms and molecules which combine to form suns and associated planets on which life evolves. Life also exhibits patterns of increasing complexity with simple organisms getting more complex over evolutionary time until they develop rationality and complex cultures.

One last note: Sometime after WWII I knew that I was going to be an engineer and fly a military aircraft. I had no idea how I was going to accomplish these two goals but I knew

---

[1] From Smith, Kelly, Philosopher and Evolutionary biologist, Clemson University, January 2015

failure was not an option. I did both by my mid twenties, which left the rest of my careers without constraints. This book presents a small sample of this knowledge.

My hope is that narratives in this book will stimulate someone's interest and cause him/her to get involved in science. No age limits apply.

My only unfilled goal is that every high school diploma should require the taking of one unit of physics!

THE ADVANCEMENT OF TECHNOLOGY: HOW OVER TIME IT EFFECTS YOUR CAREER

The advancement of technology increases by geometric progression that is defined by being relative to a point in time without specific limits of magnitude assigned to the values of measurement. For example- Technology advancement from 1950 through 1975, a mere 25 years, was greater than all time before is a reasonable assumption. Because of the societal impacts of this new technology being adopted by the engineering community, the result has been the catalyst for a technological revolution. Since the assumption is that these new technologies will be eagerly embraced by the engineers, the impact of the revolution is gauged by the lack of communication.

The technological impact of computers is an example as to how most of society and some engineers, particularly the older ones that went through engineering school with a slide rule, have resisted change.

The technical competency curve (TC) is plotted with the accumulation of TC as the "y" axis and time as the "x" axis. Time can be presented in years or as the career time of an engineer. The slope of the TC curve for any engineering discipline is increased as new formulae and solution techniques and methodology are introduced. Also from inference any device or tool design to iterate various relationships rapidly, thus provides ways to solve detailed relationships, such as finite models, economical and practically for useful applications to define a: real thing" will have an enormous impact on the TC slope. The slide rule, mechanical electrical calculator and the digital and analog electronic computers have had a tremendous effect upon mathematical computations.

The following graph demonstrates the relative comparison of the advancement of technology or state-of-the-art trend of any engineering discipline when plotted against (x-axis) an individual's competency progression. The engineer is trained (college) in a specific field, however, as his career progresses he receives little or no training. The experience factor is short lived ($\leq$ 10 years) and he becomes less effective with time. The steeper the TC curve the shorter the effectiveness of the experience curve. Rapid technological advancements cause revolutions which result in the individual's expertise fading.

TC curve number two (2) presents the "training effect" upon ones competency.

TC curve number three (3) presents the accumulated training effect. When the "refresher" training is strategically place it will keep an individual component in his career field.

TC curve number four (4) presents the effect of tool such as calculators. These devices will increase the slope of the TC curve making it imperative that the individual accelerate his/her refresher training.

As an example, using the aforementioned curve number three as a general guide and the following Exhibits, I'll present a chronological list of my knowledge linage from the beginning of my formal education. Following the physics and chemistry courses at Mullens High School, Exhibit A presents most of my knowledge gleaned from my formal education. Exhibit B is a list of engineering and/or scientific books, use as references that I have analyzed in the last 15+ years. These books reside in my library.

The most difficult book for me to understand was Stephen Hawking' "A Briefer History of Time". I read the first three chapter three times and had no idea how he presented his material. I set the book aside and after reading about 25 other books I decided to re-read Hawking' book. I understood his points and was so enthused that I read the book three times. Great read!

Normally when I have a difficult time "getting" into a technical book I try to find a chapter that has a subject that I am acquainted with and/or can comprehend, I'll read that chapter. Coupled with the background knowledge of the author I can finish the book.

The other half of my trail threads through experiences, responsibilities, and most of all through folks whose world has overlapped my world in the course of a career.

COMPUTATION SHEET

CORPS OF ENGINEERS, U.S. ARMY
OHIO RIVER DIVISION

T-C CURVE

MINIMAL TRAINING

TECHNICAL COMPETENCY

SPECIFIC

ENGINEERING

DISCIPLINE

STATE-OF-THE-ART

PROGRESSION

CURVE

INDIVIDUAL

COMPETENCY

PROGRESSION

CURVE

COLLEGE   CAREER   RETIRE

TIME

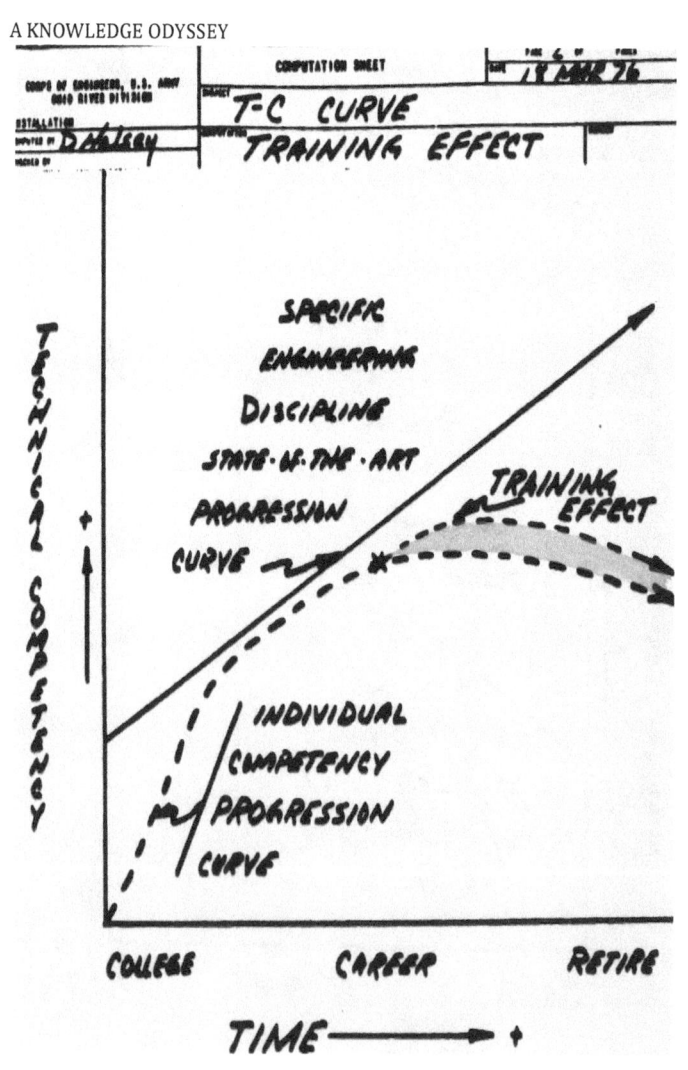

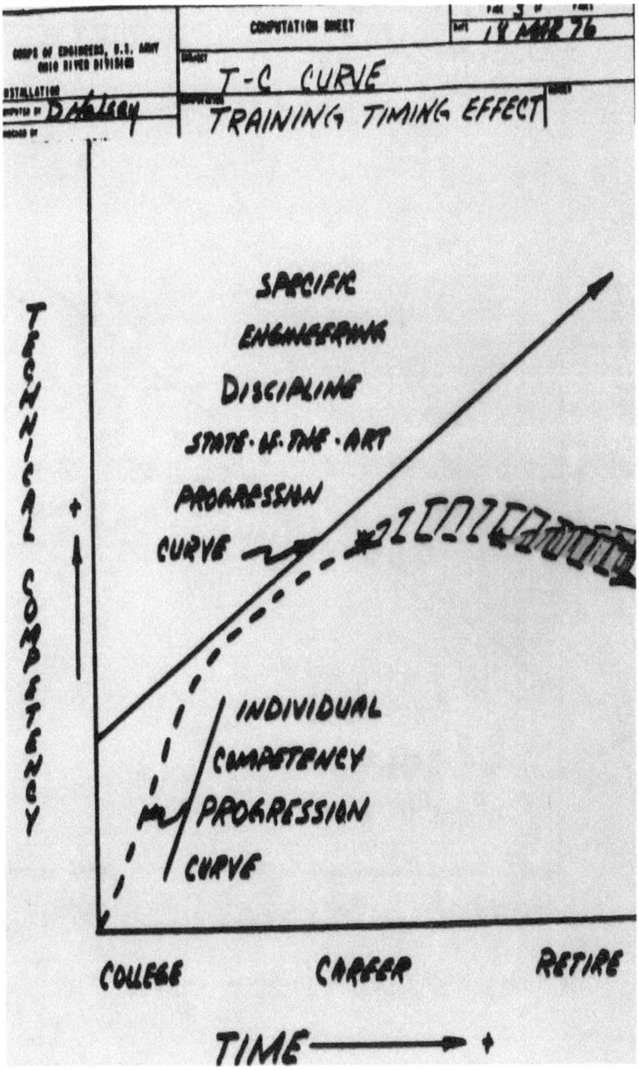
COMPUTATION SHEET
T-C CURVE
TRAINING TIMING EFFECT
SPECIFIC ENGINEERING DISCIPLINE STATE-OF-THE-ART PROGRESSION CURVE
INDIVIDUAL COMPETENCY PROGRESSION CURVE
TECHNICAL COMPETENCY
COLLEGE
CAREER
RETIRE
TIME

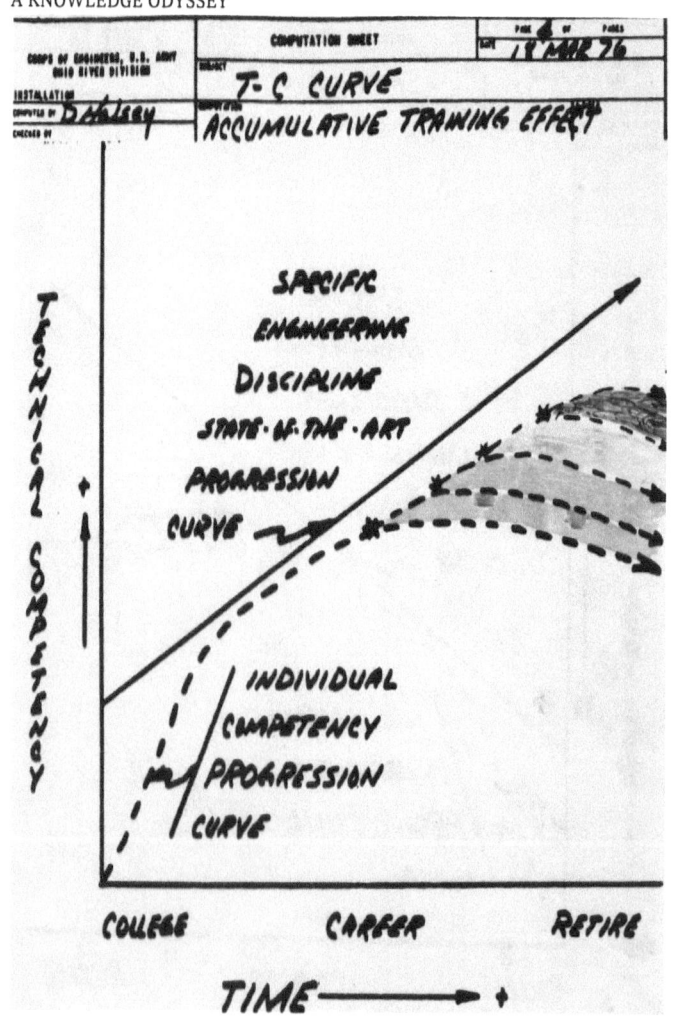

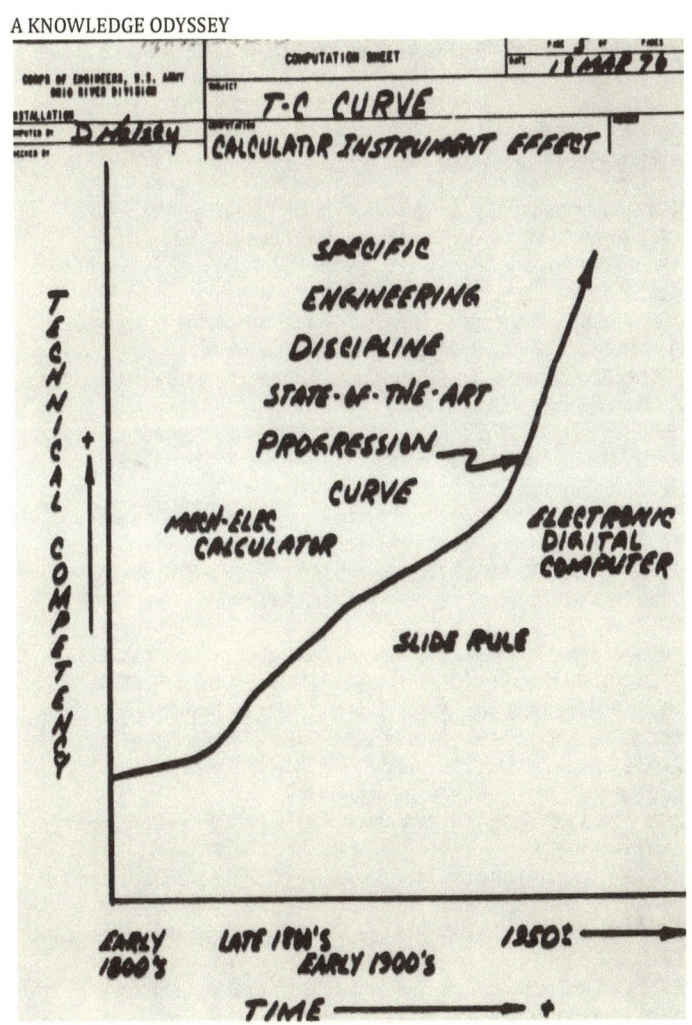

A KNOWLEDGE ODYSSEY

MY FORMAL KNOWLEDGE ENDEAVORS

Men, forever tempted to lift the veil of the future--with the aid of computers or horoscopes or the intestines of sacrificial animals--have a worse record to show in these "sciences" than in almost any scientific endeavor. Hannah Arendt

1948-Junior Nature Camp, Oglebay Institute, Wheeling, WV

1949-Knight of the Golden Horseshoe, West Virginia History

1959-*Field Artillery Officer Basic Course, U.S. Army Artillery and Missile School, Ft Sill, OK

1960-*Primary Flight and Academic Courses, Hawthorne School of Aeronautics, U.S. Army Aviation School, Ft Rucker, AL

1960-*Officer Fixed Wing Aviator Course, Phase B (Tactical), U.S. Army Aviation School, Ft Rucker, AL

1960-*Officer Fixed Wing Aviation Course, Phase C (Instruments), U.S. Army Aviation School, Ft Rucker, AL

1961 -*Safety and Command, Ft Meade, MD

1962 -*Altitude chamber training, Naval Air Station Patuxent River, VA

1962 -*Ejection seat training, Naval Air Station Patuxent River, VA

1965-*Armor Officer Basic Course, U.S. Armor School, Ft Knox, KY

1966-Hydraulic Engineering Theory, Graduate School, University of Cincinnati, OH

1966-Air and Water Pollution, Engineering Society of Cincinnati, OH

1966-*CS Perception, U.S. Army Chemical Corps School, Ft McClellan, AL

1966-*Public Health and Medical C and B Defense, Department of Health, Education, and Welfare, Public Health Service, Ft McClellan, AL

1966-Storage-Yield Relations and Stream flow Simulations, Hydrologic Engineering Center, Department of the Army, CA

1967-Value Engineering Indoctrination, U.S. Army Corps of Engineers, Cincinnati, OH

1968-Personnel Management for Executives Conference, Department of the Army, Ft Benjamin Harrison, IN

1971-The Effective Executive, U.S. Army Corps of Engineers, Cincinnati, OH

1972-Public Speaking, U.S. Civil Service Commission, Cincinnati, OH

1975 -Astronomy, Thomas Moore College, KY

A KNOWLEDGE ODYSSEY

1975-*Civil Disturbance Orientation Course, U.S. Military Police School, Ft Gordon, GA

1975-EEO Indoctrination, U.S. Army Corps of Engineers, Cincinnati, OH

1975-*Engineer Officer Advanced Course, U.S. Army Engineer School, Ft Belvoir, VA

1975-Understanding Human Behavior (Transactional Analysis), Xavier University, Cincinnati, OH

1976-Installation and Inspection of Hydraulic Turbines, Allis Chalmers Corp., York, PA

1976-Executive Development Assignment detail, Acting Assistant Chief of Engineering Division, Louisville District, US Army Corps of Engineers, Louisville, KY

1977-Factor Evaluation System, U.S. Civil Service Commission, Cincinnati, OH

1978-The Art of Negotiating Course, Cincinnati, OH

1978-Time Management Seminar, Xavier University, Cincinnati, OH

1979-*Command and General Staff College, U.S, Army Command and General Staff College, Ft Leavenworth, KS

1980-Citizen Participation, U .S. Army Corps of Engineers, Washington, DC

1980-Professional Engineer course preparation, Cincinnati Technical College, OH

1980-Registered Professional Engineer, State of West Virginia

1982-OSM Management Seminar, U.S. Department of the Interior, Office of Surface Mining, Berkeley Springs, WV

1982-*Reserve Components National Security Course, National Defense University, Washington, DC

1982-Washington Management Institute, Washington, DC

1983-OSM Management Seminar, U.S. Department of the Interior, Office of Surface Mining, Breckenridge, CO

1984-Coaching the Experience Driver Course, U.S. Office of Surface Mining, Herndon, VA

1984-Mine Gases, U.S. Department of Labor, Beckley, WV

1985-*Near East and North Africa Area Studies, Foreign Service Institute, U. S. Department of State, Washington, DC

1986-Professional Engineer, Virginia

1986-School of Offshore Operations, University of Texas at Austin, Houston, TX

A KNOWLEDGE ODYSSEY

1987-Aviation Week Research Advisory Panel, Aviation Week and Space Technology Magazine

1987-*Civil Affairs Officer Advanced Course, U.S. Army Special Warfare Center and School, Ft Bragg, NC

1989- Aviation Management Seminar, Office of the Secretary, DOI, Aircraft Services, Washington, DC

1990-Adult & Pediatric Heart Saver course (CPR), Fire & Rescue Department, Fairfax County, VA

1991-Total Quality Management in the Public Sector, Western Executive Seminar Center, Denver, CO

1992-Conducting Management Control Reviews, Management Concepts Inc, Vienna, VA

*Military

A KNOWLEDGE ODYSSEY
READING LIST FOR THE ADVANCEMENT OF TECHOLOGY

Of the vast array of literature on science, specifically physics, the brain and how it processes information; and selected histories of the progression of each; I have read only a small fraction. Included in this list is that portion that I have found in the last few years that I consider illuminating. I own all.

1. Abrams, Nancy Ellen; Primack, Joel R., "The View from the Center of the Universe", Riverhead, 2006
2. Aczel, Amird, "The Riddle of the Compass", Harcourt, 2001
3. Adams, Fred & Laughlin, Greg, "Five Ages of the Universe", Simon & Sch., 1999
4. *Asimov, Isaac, "Asimov's New Guide to Science", Basic Books, Inc., New York 1984*
5. Asimov, Isaac, "Atom- Journey Across the Sub Atomic Cosmos", Plume, 1992
6. Auyang, Sunny Y., "How is Quantum Field Theory Possible?", Oxford Univ., 1995
7. Barbour, Ian C., "When Science Meets Religion", Harper, San Fran., 2000
8. Barratt, Krome, "Logic & Design in Art, Sciences & Mathematics", Design Books, NY 1980
9. *Barrow, John D., "The Book of Nothing", Pantheon Books, New York 2000*
10. Bartusiak, Marcia, "Einstein's Unfinished Symphony", Henry, 2000
11. Behe, Michael J., "The Edge of Evolution", Free Press, 2007
12. Benacciho, Professor L, "The Great Atlas of the Universe", D&C, 2007
13. Berry, Adrian, "Galileo & the Dolphins", Wiley, 1996
14. Bloom, Howard, "Global Brain", Wiley & Sons, 2000
15. Boslough, John; Mather, John, "The Very First Light", BasicBooks, 1996
16. Boss, Alan, "Looking for Earths", John Wiley, 1998
17. Bova, Ben, "Faint Echoes, Distant Stars", HarperCollins, 2004
18. Brockelman, Paul, "Cosmology and Creation", Oxford University Press, 1999
19. Brooks, Michael, "13 Things that don't make sense", Doubleday, 2008
20. Bruce, Colin, "Schrodinger's Rabbits", Joseph Henry Press, Washington, D. C., 2004

21. Bryson, Bill, "A Short History of Nearly Everything", Broadway Books, Random House, 2003
22. Buchanan, Mark, "Ubiquity", Crown Publishers, NY 2001
23. Caes, Charles J., "Cosmology", Tab Books, Inc., Blue Ridge Summit, PA, 1986
24. Casti, John L., "Paradigms Lost", Avon Books, NY, 1989
25. Casti, John L., "Would be Worlds", John Wiley, 1997
26. Chown, Marcus, "The Quantum Zoo", Joseph Henry Press, 2006
27. Cole, K. C., "Mind over Matter", Harbour, 2001
28. Cole, K.C., "The Hole in the Universe", Harcourt, Inc., NY 2001
29. Darling, David, "Teleportation", Wiley, 2005
30. Davies, Paul, :About Time", Touchstone Books, 1995
31. Davies, Paul, :The 5th Miracle", Touchstone Book, London, 1999
32. Davies, Paul, "The Goldilocks Enigma", Mariner, 2006
33. Davies, Paul, "The Last 3 Minutes", Basic Books, 1994
34. Dawkins, Richard, "The God Delusion", First Mariner, 2008
35. Doidge, Norman, M.D., "The Brain that Changes Itself", Viking, 2007*
36. Drake, Stillman (Translator), "Discoveries & Opinions of Galileo", Randon House, 1957
37. Edelman, Gerard M., "Bright Air, Brilliant Fire", Basic Books, 2005
38. Ferguson, Kitty, "Measuring the Universe", Walker, 1999
39. Ferris, Timothy, "Coming of Age in the Milky Way", HarperCollins, NY, 1988
40. Ferris, Timothy, "The Science of Liberty", HarperCollins, 2010
41. Ferris, Timothy, "The Whole Shebang", Simon & Sch., 1997
42. Flatow, Ira, "Present at the Future", HarperCollins. 2007
43. Freeman, Ken & McNamara, "In Search of Dark Matter", Praxis, 2006
44. Gingerich, Owen, "The Book Nobody Read", Penguin Books, England, 2004
45. Gleick, James, "CHAOS", Penquin Books, 1987
46. Gleiser, Marcelo, "The Dancing Universe", Plume, 1998
47. Goldblum, Naomi, "The Brain-Shaped Mind", Cambridge Press, UK, 2001
48. Goldsmith, Donald, "The Runaway Universe", Perseus, 2000

49. Gott III, J. Richard, *"Time Travel in Einstein's Universe"*, *Houghton Mifflin Co., Boston, 2001*

50. Green, Brian, *"The Elegant Universe", Vintage Books, New York 1999*

51. Gribbin, John, "In Search of the Big Bang", Penguin Books, 1998

52. Guth, Alan, "The Inflationary Universe", Perseus, Cambridge, MA 1997

53. Hawking, Stephen and Mlodinow, Leonard, "A Briefer History of Time", Bantam Dell, New York, 2005

54. Hawking, Stephen, "Black Holes and Baby Universes and Other Essays", Bantam Books, London 1993

55. Hawking, Stephen, "The Universe in a Nutshell", Bantam Books, London 2001

56. Hawkins, Stephen, "A Brief History of Time", Bantam Books, London, 1988

57. Hawkins, Stephen, "On the Shoulders of Giants", Running Press, Philadelphia, PA, 2002

58. Heudin, Jean-Claude (Edited), "Virtual Worlds", Westview, 2004

59. Hogan, James P., "Kicking the Sacred Cow", Simon & Schuester, 2004

60. Holmes, Hannah, "The Secret Life of Dust", John Wiley & Sons, 2001

61. Hooper, Judith, "Of Moths & Men", Norton, 2002

62. Illingworth, Valerie, *"The Facts on File Dictionary of Astronomy", Facts on File Publications, New York 1985*

63. Kaku, Michio, "Beyond Einstein", Anchor Books, 1995

64. Kane, Gordon, *"Supersymmetry", Helix Books, Cambridge, MA 2000*

65. Klein, Etienne; Lachieze-Rey, Marc, "The Quest for Unity", Oxford University Press, 1999

66. Kragh, Helge, "Cosmology & Controversy, "Princeton Press, 1996

67. Kruszelnicki, Karl, "Great Myth-Conceptions". Andrews McMeel, 2006

68. Laughlin, Robert B., "A Different Universe", Basic Books, New York, 2005

69. Lindley. David, "Uncertainty", Doubleday, 2007

70. Lloyd, Seth, "Programming the Universe", Alfred A. Knopf, 2006

71. Loewenstein, Werner R., "The Touchstone of Life", Oxford, 1999
72. Maddox, John, "What Remains to be Discovered", The Free Press, NY, 1998
73. *Man, John, "The Encyclopedia of Space Travel & Astronomy", Octopus Books Limited, London 1979*
74. Mazur, Joseph, "The Motion Paradox", Dutton, 2007
75. Miller, Arthur I., "Empire of the Stars", Hpughton Mifflin, 2005
76. Minsky, Marvin, "The Society of Mind", Simon & Schuster, 1985
77. Moffett, Shannon, "The Three-Pound Enigma", Algonquin, 2006
78. Moffett, Shannon, "The Three-Pound Enigma", Algonquin, 2006
79. *Moore, Patrick, "The Picture History of Astronomy", Gosset & Dunlap, New York 1972*
80. *Moring, Gary F., "The Complete Idiot's Guide to Theories of the Universe", ALPHA A Person Education Co., 2002*
81. Morrison, Philip & Phylis, "The Ring of Truth", Randon House, 1987
82. Ochoa, George; Hoffman, Jennifer; Tin, Tina; "Climate", Rodale, 2005
83. Park, David, "The Fire Within the Eye", Princeton University Press, 1997
84. Parker, Barry, "Creation-The Story of the Origin and Evolution of the Universe", Perseus Books, 1988
85. Penrose, Roger, " The Road to Reality", Alfred A. Knopf, 2004
86. Penrose, Roger, "The Emperor's New Mind", Oxford University Press, 1989
87. Pinker, Steve, "the blank slate", Viking, NY, 2002
88. Pollack, Robert, "The Missing Moment", Houghton Mifflin Co., 1999
89. Powell, James Lawrence, "Mysteries of Terra Firma", The Free Press, NY, 2001
90. Primack, Joel R., Abrams, Nancy Ellen, "The View from the Center of the Universe", Riverhead, 2006
91. Rees, Martin, "Just Six Numbers", Basic Books, 2000
92. Rizzi, Anthony, "The Science Before Science", IAP Press, Baton Rouge, LA 2004

93. Robinson, Andrew, The Last Man Who Knew Everything", PI, 2006

94. Roston, Eric, "The Carbon Age", Walker, 2008*

95. Schwartz, Jeffrey M. and Begley, Sharon, "The Mind & The Brain", HarperCollins, NY 2002

96. *Siegfried, Tom, "The Bit and the Pendulum", John Wiley & Sons, Inc., New York 2000*

97. Silver, Brian, "The Ascent of Science", Solomon, 1998

98. Smolin, Lee, "The Trouble With Physics", First Mariner Books, 2006

99. Steinhardt, Paul J., Turok, Neil, "Endless Universe", Doubleday, 2007

100. Strogatz, Steven, "Sync", Hyperion, NY, 2003

101. Swimme, Brian & Berry, Thomas, "The Universe Story", Harper San Francisco,1992

102. Warshofsky, Fred, "Stealing Time", TV Books, 1999

103. Weinberg, Steven, "The First Three Minutes", Perseus Books, 1977

104. Wilson, Edward O., "The Future of Life", Vintage Books, 2002

105. Woit, Peter, "Not Even Wrong", Basic, 2006

106. Wolf, Fred Alan, "Parallel Universes", Touchstone, Simon & Schuster, NY, 1988

107. Yourgrau, Palle, "A World Without Time", Basic Books, 2005

108. Hawking, Stephen & Mlodinow, Leonard, "The Grand Design", Bantam, 2010

109. Carroll, Sean, "From Eternity to Here", Dutton, 2010

TRAINING ENGINEERS FOR USING AND MANAGING COMPUTERS

Act as if you were going to live forever and cast your plans way ahead. By this I mean that you must feel responsible without time limitation; and the consideration whether you may or may not be around to see the results should never enter your thoughts. If your contribution has been vital ,there will always be somebody to pick up where you left off and that will be your claim to immortality" Reginald A. Hubley

Introduction

Before 1972, engineer managers remained completely aloof from any utilization of automatic data processing (ADP) and engineering applications. Everybody went through engineering school using a slide rule. ADP folk were left to decide what engineering requirements were and what computer hardware was necessary to fulfill that requirement.

The Ohio River Division was one of the first Corps of Engineers' Divisions to have a computer. At that time (1965) engineers (Water Resources) procured the computer and ran its operations. The computer was used effectively to crunch numbers in real time operation of water resources.

Once the computer concept began to grow within the Corps, ADP centers were established, thus separating operator from user. This caused a need for communication and coordination between the two groups. A need that progressed very slowly. Finally, the Office of the Chief of Engineers, Army Corps of Engineers, Washington, DC (OCE), realized that a void between ADP and ADP practitioners existed and was growing at an alarming rate. Therefore, something had to be done if the Corps ever expected to keep current with the state-of-the-art in technical computer applications. Each District and Division were continually "reinventing the wheel" in almost every application.

OCE began publishing abstracts of all applications' programs; however, this approach did not work for various reasons. The theory being that an engineer could look at the list first for a needed program. One major problem with this theory was that several programs had different names. Yet they were basically the same, so the list grew beyond controllable numbers.

Finally, in September 1968, OCE created the "Engineer Computer Concepts and Application Group"[2] (ECCAG) as a continuing body composed of carefully selected engineers and computer specialists of the Corps having expert knowledge of computer-assisted engineering applications. The Group functioned under the direction of the Chief, Engineer Information and Data Systems Office (CEIDSO), OCE. Members of the Group were appointed by CEIDSO from a list of nominees submitted annually by the Divisions, Districts and other Class II installations and activities. Each membership term was three years. In addition to the ECCAG, the Director of Civil Works (DCW) would designate program review monitors in each engineering specialty in Civil Works who had full knowledge of pertinent existing or purposed computer applications.

Objectives- The ECCAG initially had three general objectives: (1) To analyze Corps-wide requirements for effective use of the computer as an engineering tool by evaluating existing and proposed practices developed by various field offices for different applications; (2) To formulate guidelines for, and prescribe as desirable and practicable, standardization of computer hardware and software for Corps-wide application to engineering applications particularly with respect to centralized computer systems and uniform or compatible computer languages and programs.; (3) To familiarize the engineer with existing programs which apply to specific design methods or a problem and to permit his use of these programs with a minimum amount of effort, if he/she so desires. This objective did not imply that the engineer's professional integrity was to be compromised, nor was it intended to limit the engineer's choice of options for problem solutions.

------

[2] *This author was a charter member, representing the Ohio River Division, first being the Chairman of the Training Committee (two years) and then Chairman of the Group (five years)*

The ECCAG had one major obstacle to overcome. In 1968 every Engineer, from executive to recent graduates, in every position, both civilian and probably military, went through engineering school with a "slide rule"[3]! If the Corps was to maintain its position as one of the world's top engineering organizations, it had to change the computer culture quickly. Training courses were organized based upon the recommendations of the ECCAG to indoctrinate the "slide rule" engineers at all levels of responsibility.

The courses were presented near most Corps Division offices but at remote locations e.g. State Parks, etc. The instructional staff for each course was gathered from a representative cross-section of organizations that was interested in and involved in the use of computers in engineering applications. Their expertise, experience, interests, and opinions represented, to the greatest extent possible, the full range of thought in most important aspects of computer applications in engineering. Represented were computer gurus, familiar with engineering applications, from academia, technical labs, in-house and sister federal agencies.

The following is from the lecture entitled "**Training Engineers for Using & Managing Computer-Studies**" that this author presented in March 1976 at Portland, Oregon.

"Present were engineering executives who had little knowledge in computer capabilities that existed in the 1970s. NOTE: This lecture site was the first time that I introduced the "Knowledge" curves contained in chapter 1 as a trial balloon to get a reaction. The response was excellent but I never expanded the thoughts until now.

Introduction- *One of the most frustrating experiences that this author had as an engineer began about 1958 and it hasn't ended yet (2015). It's simply "why don't we as practicing engineers continue to*

---

[3] *As did this author*

improve or accept or adapt the state-of-the-art techniques, tools, methodology, etc., to solve our technical problems? Why aren't we constantly searching for better technical ways to accomplish our jobs? Why do we continue to become embedded in traditional problem solving techniques and lose, to some degree, our intense or eager interest and enthusiasm which are prerequisites for success in our respective technical fields? Why aren't we striving to improve our efficiency to the point where today we can do, by several fold, the amount of technical computation, faster, less costly, more efficiently, and in more technical detail and precision than a few years ago? Not an impossible task for government oriented engineers.

During the late 1950's and 1960's, a group of engineers and scientists working for a government entity called NASA proved that any conceivable technical problem could be solved from conception and derivation of a solution methodology to a real-time application.

How do we motivate ourselves to apply computers in solving our technical problems with sophisticated, relatively new techniques which are only possible with a computer? How is self-confidence built in a design or any other application when we are not sure what transpires between computer input and output?

Training courses, such as this one, may only whet your appetite or stir your heretofore dormant will to gain knowledge in the realm of the technically unknown. Some of you are probably thinking that the specific engineering computer applications that are presented during this week are research-and development-oriented, theoretical, or ivory tower approaches to problem solving and are still, for the most part, in the laboratory stage. Let me assure you that this is not the case. The majority of the applications that will be presented were developed to solve practical engineering problems similar to your own. Most of the procedures and methods presented are continuously being improved with every application. However, in most cases, the base methods are several years old.

A KNOWLEDGE ODYSSEY

So the question still exists; how do we motivate you?

Currently, the most successful techniques of improving engineering computer application have been by "command force". Various committees, such as ECCAG have been very successful because of "command force". The use of Engineering Regulations, such as the one requiring all engineers and scientists to receive computer training commensurate with their respective positions is an example of "command force". Without this force you probably would not have "volunteered" for this course or any other course relating to engineering computer applications. The press of your job or time schedules for meetings such objectives as milestones would have prevented you from even thinking about this course had it not been for a Regulation or "command force".

Throughout this course, you will see and hear from the "top-of-the-line" in engineering computer applications. Folk from the academic, private and government worlds will present specific engineering applications and philosophies developed to solve real problems. You should be able to communicate, understand, and probably wonder why you cannot apply the same or similar techniques to solving your own real problems. Following this course, you should realize that computers and ADP practitioners don't speak an unknown tongue. You can and will communicate with these instructors and understand what they say. Hopefully, this will erase from your minds part of the fear of computers and you can begin to see how utterly simple it is to become involved.

The ultimate success of any training endeavor depends upon what happens after the course has been taught. What is done in follow up? Too many courses are attended with great enthusiasm on the part of the attendee. However, when he/she returns to his/her home station, they could become technically frustrated because their supervisor continues to push traditional problem solving techniques because of deadlines or other pressing reasons. Other tolerances for the final product have been broadened to the extent that negates more precision. Design, formulation plans, etc., are controlled by time, money and not necessarily by comprehensive evaluation.

On the other hand, if attendees come home with the problem solving technique which so happens to be the very technique that will solve an existing problem, and he/she so demonstrates this to their supervisor, then this will become the accepted way in that office.

We have difficulty in foreseeing savings in future jobs by changing methodology on the current job. Time, money and person-power constraints imposed by the organization have, in some instances, become a deterrent to the advancement of the technical elements.

In the past, the engineers and scientists was conceivably the corral fence around the administrative elements that were the so-called horses which we surrounded; they supported us! Now just the opposite occurred. The engineers or the technical elements were the horses surrounded by the administrative staff that became the corral or the governing force. This was the new reality.

Probably the best available training technique for advancing engineering computer applications is actual application or demonstration, using numerical solutions techniques to solve specific problems for individual engineers on a one-on-one basis. Based on my involvement in using computers for solving real time problems pertinent to water control management activities, I feel that the most enthusiasm and use among engineers has developed from having the computer equipment available (seeing the blinking lights, pushing the buttons, etc.) where engineers can have hands on than from any other training vehicle.

Training Engineers in General- Performance efficiency, particularly in an environment of computer-aided engineering practices, is directly related to training accomplishments. Therefore, training engineers for using and managing computer studies is the keystone *to success in an environment of computer-aided engineering applications. Supervisors are responsible to and for your subordinates' capabilities to perform their respective technical functions in a manner that is commensurate with the state-of-the-art for their particular engineering discipline and/or applications.*

Training is vital to the health of any technical organization. In most instances new ideas, concepts and techniques must flow from outside an organization. Without new ideas from outside an entity becomes stale. Self taught entities have built a "knowledge fence" around themselves, thus the advancement of ideals ceases. This is particularly true for entities where technical production is measured in quantitative and qualitative terms. Technical folk must receive periodic training in carefully selected applicable technical areas. The accomplishment of this awesome task is the responsibility of the technical supervisor at every level.

During the decade of the 1960s the computer made a tremendous impact upon the educational processes. College engineering curriculums require that students be proficient in computer applications relative to their respective field of study.

Parallel to the educational changes are the existing approaches to engineering problem solving being made by small groups of individuals in private and government. These progressive engineers are successfully applying dynamic problem solving techniques to practical problems. The mathematics employed in these approaches can only be attempted with the aid of a computer. As an end result, existing problems are being analyzed in more detail with more comprehensive solutions than ever before. In the past one could only provide one answer in the time allotted, but, with a computer, multiple solutions can be had in a fraction of the manual time.

Training selected individuals in specific technical applications appears to be the most practical way to spread the knowledge. One caution when training, you need a backup if one person leaves the organization, so educate multiple people based upon the importance of the knowledge to the organization.

History (1967-1980) of the growth of daily computer applications in water resource management[4], Reservoir Regulation and the Water Quality Sections

Reservoir Regulation Section: The "bread and butter" application (1976) is a mathematical model of the Ohio River main stem (981 miles) which is used: (1) to assist in regulation of reservoirs for maximization of operational benefits where Ohio River controls exist as part of a reservoir authorized objective; (2) to assist in lowering flood stages along the lower Mississippi River by operating Barkley (Corps) and Kentucky (TVA) reservoirs to reduce flood crests at Cairo (confluence of Ohio and Mississippi Rivers); (3) to make operational forecasts of river stages at cofferdams for new high lift[5] navigational dams now being constructed along the Ohio River; (4) to assist in scheduled maintenance at low lift navigation dams; (5) to develop gate opening schedules for new navigation dams to ensure a steady movement of water during low flow to maximize reaeration (restoring oxygen to water); and (6) to make flow and stage forecast for any segment for emergency situations.

The above model replaced a manual technique which required three engineers, working about one and one half hours to route (using a calculator or slide rule) the flow only for the lower half of the main stem. Currently (1976) the aforementioned model takes about 10 minutes to run, including the curve (hydrograph) plotting, flow adjusting, etc.

---

[4] *Author was chief of Reservoir Regulation, assistance chief of the Corps Ohio River Division's Reservoir Control Center, and Engineering Computer coordinator from 1965 to 1980.*
[5] *There are 19 high lift dams (tainter gates) replacing 52 low lift dams (wickets)*

A more sophisticated model to replace the above one was developed by the Mathematical Hydraulics Division at the Corps Engineer Research and Development Center, Vicksburg, MS. The model employs the numerical solutions of the unsteady flow equations, continuity and momentum. The model system, composed of portions of the lower Ohio, Cumberland, Tennessee and Mississippi Rivers, is currently operational and is being used daily. This model has the capability to handle any system with an unlimited number of junctions or tributaries. It contains a tremendous amount of channel geometric data, roughly five mile increments, and computes flow using five minute increments. This small segment of the model takes about 30 minutes to run on a GE 600 computer and about five to eight minutes on a UNIVAC 1108 system. The model takes a very small amount of input and only produces that part of the output which the user desires. Downstream stages can be accurately forecast to within a half tenth of a foot, whereas, before the accuracy was plus or minus one half foot, plus it took about an hour per computation, manually. Because of the tremendous amount of channel geometric data used by the model, it is accurate on an hourly basis, particularly when making discharge changes from Barkley and Kentucky Dams. Further development of this model[6] for the entire Ohio River is in progress.

Several bookkeeping programs are used daily to provide a status report of the 72 (1976) Corps' multiple-purposed reservoirs, within the Ohio Basin (202,000+ square miles), to maintain statical data on rainfall and flow at critical locations. By inputting the current pool elevations for each project, the program computes an analysis for each project, each tributary system, and for the entire Ohio Basin system. This program takes about five minutes to run.

Other computer applications involved a program to route reservoir holdouts (storage increases within the reservoir pool thus prevented

---

[6] *WES technical Technical Report N-74-8 contains a detailed description of this model.*

from adding to the downstream flooding) for computation of reservoir benefit analysis. Normally 12 flood periods are routed through a proposed reservoir project to determine the downstream benefits. Manual computations took about four and one half days per reservoir. The computer model took less than 10 minutes per reservoir.

There are numerous other statistical analysis programs that involve analyzing hydro met (hydrology and meteorology) and chemical data on a daily basis and continually making comparisons with weekly, monthly and yearly averages.

Currently Purdue University Hydraulics Department personnel (Corps supporting the hydraulics doctoral program) are constructing several models for use in solving the problems of estimating side flow into reservoir, river-river systems routing, reservoir operations simulation, forecasting and optimizations. The PhD candidates are doing a Corps agreed dissertation by modeling an Ohio tributary system with the idea that these models will be used in daily water management activities. These are quantity models. The Purdue effort has completed two years (1976) of a five year program.

Water Quality Section: *During the last four years (1972-76) computer applications have grown. The application of temperature (thermal) models to create a heat budget analysis for proposed lakes is being made for the purpose of designing multilevel intake structures. This knowledge will be used to create a two story fishery in thermal stratified lakes, as well as to control outflow temperatures. Also models have been applied to existing lakes to determine the reservoir hydrodynamics so that reservoir outflow temperatures can be anticipated.*

These simulations models require a large computer to run effectively. In order to run these models, a tremendous amount of hydrology and meteorology data, such as a definition for sunlight, dew point and water temperatures are needed.

There are various other modeling applications which will eliminate more of the manual effort. One word of cautioned that has never left our minds; make sure you can function if the power goes off! Water flow will not stop.

Summary
Most engineering computer applications require basically the same steps; collect data to satisfy an empirical relationship which simulates or approximates an object or state. ***Make sure that the interpretation of the answer does not go beyond the significant data input.***

WATER RESOURCE MANAGEMENT-AN ENGINEER'S NEW FRONTIER

Water is the driver of Nature. Leonardo da Vinci

We forget that the water cycle and the life cycle are one.- Jacques Cousteau

INTRODUCTION

The U.S. Army Corps of Engineers experienced a revolution in water resource management, specifically dam building, when the National Environmental Policy Act of 1969 (Public Law 91-190) (NEPA) and the Rivers and Harbor Flood Control Act of 1970 (Public Law 91-611) were signed into law. The Corps was blindsided by the new environmental studies that a dam project needed for passage by Congress. Environmental policy which the Corps was blamed for not following had, in fact, never existed. New guidelines for environmental statements for every project had to be developed, as well as social impacts such as disruption to a community, etc.

The Corps developed a comprehensive intra-agency, local, state and federal strategy for solving these new problems. The ***Reservoir Regulation Complexities***[7] chart depicts these relationships.

In summary, the Corps quickly increased staff expertise to go beyond the requirements of these two laws. After more than 40 years every dam project had exceeded its design benefits for every authorized purpose including flood control, low flow, to recreational. In fact all of the projects accumulated more flood control benefit than the project's cost to build within their first decade of operation!

This revolution made the Corps of Engineers a much better technological organization. In fact, it appeared to be the most

---

[7] *I develop while Chief of reservoir Regulation for the Ohio River Division.*

27

advanced environmental-water resource-engineering organization on the planet!

The following was taken from a paper written for the "*Military Review*" to satisfy requirements for this author's graduation from the U.S. Army's Command and General Staff College. The paper is dated 24 February 1977.

Since 1824, the Corps of Engineers has been the primary developer of this country's water resources. The Flood control Act of 1936 added the responsibility of constructing multiple purpose reservoir projects nationwide. It wasn't until after W.W. II that the design and construction of flood control reservoirs got into "hi gear". By the early 1970's, the Corps had completed over 350 reservoirs thus creating a monumental water management task for the respective Corps office.

During Lieutenant General F. J. Clark's tenure as Chief of Engineers, he outlined, by letter dated 6 August 1970, the Corps' water management philosophy. He said in part: "The need for emphasis on effective reservoir management continues to grow as projects increase in numbers and complexity and demands for services multiply. Flood control, navigation, and hydroelectric power still represent basic services, while recreation, water quality, water supply and conservation activities in general have become major goals."

"The impact of reservoirs on environmental conditions and water quality control is receiving national attention. The fact that reservoirs provide a means of enhancing environmental conditions and improving water quality in streams is not fully recognized by the public, and such influences still need more precise evaluation by responsible authorities. Plans and operating techniques to alleviate or avoid existing or potentially adverse effects of reservoirs on water quality and environmental conditions require continuing study and implementation. A balanced perspective must be maintained in developing reservoir regulation plans to meet all project objectives as

effectively as possible under prevailing circumstances. The Corps of Engineers has an opportunity to play a leadership role in these activities."

Therefore, the Civil Works side of the Corps of Engineers (military is the other side) must expand the technical knowledge of its water control organization, beyond planning and construction, to fulfill this water management philosophy. In order to place this added responsibility in perspective a detail discussion of the water resource management (reservoir regulation) complexities, to include computer applications will be presented. Also, an example of seemingly simple reservoir operations will be presented to demonstrate or reiterate the complexities of daily reservoir regulation involvement by the applicable Chief of the Reservoir Regulation for the Division Office. This example follows:

RESERVOIR REGULATION COMPLEXITIES
While reservoirs are normally authorized for one or more primary purpose, secondary or minor project functions can sometimes become the dominate purpose at any given period of project life. Public awareness has demanded that each project be managed as a multiple purpose reservoir, regardless of the authorization document.

As complex as designing and constructing a dam can be, the solution of the social, economical, political, and ecological problems that arise before, during, and after a dam is built may be far more difficult. Each problem is unique; therefore, continued coordination and data exchange with many agencies and organizations that have an interest, implied either by law or otherwise, is necessary.

A partial list of such agencies or groups are: National Weather Service, U.S. Geological Survey, Bureau of Reclamation, Federal Power Commission, Environmental Protection Agency, Fish and Wildlife Commission, Office of Emergency Preparedness, State and local government's Basin Commissions, Izaak Walton League, Sierra Club, National Audubon Society, Conservancy Districts, Coordinating

Committees, and many local organizations and individuals. Each may have a bona fide concern for a specific problem(s) but none is responsible for evaluating the entire reservoir regulation sphere except the Corps of Engineers.

The chart on **Reservoir Regulation Complexities** presents a graphical representation of the primary problem areas (authorized purposes) and their interrelationship with specific application areas. The technical fields involved in problem solving are also shown. These are some of the disciplines which the engineer must be capable of molding into a functional multiple-disciplined staff.

RESERVOIR DRAWDOWN COMPLEXITIES
The following presents the phenomenon of reservoir drawdown which reiterates the complexities of reservoir regulation, as well as describe the routine problems facing any reservoir regulation person.

INTRODUCTION-Reservoir drawdown involves the entire realm of reservoir regulation. The problems are complex because they cover all authorized reservoir purposes, including changes and modifications in those purposes as design concepts become realities. Therefore, the reasons for drawing down reservoir pools are not always apparent to all interested parties. Rules and regulations for reservoir operation are based on hindsight, initially. However, after a project has existed for a while, the demands, resulting from the changing attitudes of local, regional and national interest, will usually alter reservoir operation guides. This phenomenon makes an attempt to correlate the magnitudes (volume in terms of acre feet; pool elevation in terms of feet; and time in terms of seasons, etc.) of drawdown for any two projects or system of projects extremely difficult.

DEPLETION REASONS-(A) Reservoirs are drawn down during various times for several primary reasons. They are: (1) flood control; (2) hydropower; (3) low flow (augment downstream flows): (a) water quality; (b) navigation; (c) water supply; (d) fish and wildlife;

(4) maintenance and repairs; (5) emergency; (6) aquatic controls; (7) combination of above.

The following is a summary/explanation discussion of the aforementioned reasons:
(1) Flood control-Drawdown associated with flood control operations is generally the least controversial of the aforementioned reasons. There are three primary considerations involving flood control drawdown operations. They are: (a) Releasing storage in advance of an impending flood, which will reserve additional reservoir storage capacity for the next flood; (b) Releasing storage accumulated during a flood; (c) Releasing seasonal pools, generally during late summer and fall, to gain a greater storage capability during the flood season. E.g. the fall drawdown of Summersville Reservoir, located on the Gauley River in the Corps' Huntington District, West Virginia, creates one of the highest rated white water courses in the world.
(2) Hydropower-Drawdown resulting from power generating operations is normally the most flexible and irritating because the pools could fluctuate daily. Since the pool elevation guide curves are zoned, anticipated elevation within the zone for any given time will be dependent on the power demands and reservoir inflows. Drawdown for hydropower normally begins in mid-summer and continues until the flood season; winter and early spring in the Atlantic and Gulf section of the U.S.
(3) Low flow- Drawdown for low flow purposes involves several factors, including water quality parameters, minimum releases, navigational releases, water supply, etc. Guide curves are generally zoned to protect the reservoir during drought years, and to reduce the releases dependent on the time of year and storage remaining. Drawdown occurs in the fall normally, in preparation for flood season. Drawdown for low flow purposes is generally the most controversial of all the reasons, because of current public attitudes toward lake recreation activities.
(4) Maintenance and repairs- Reservoirs are drawdown periodically to perform inspections on outlet works, clean, repair or build launching ramps, docks, etc. This type drawdown normally occurs at

the end of the fall drawdown or end of summer recreational activities, when the pool is at minimum levels. It is a planned and coordinated effort and normally causes little conflict with other reservoir functions.

(5) This type drawdown occurs when the integrity of the dam is questioned because of leakages, boils (water escaping the dam via tree root decay, varmint holes or a breach of integrity of the dam), malfunction of outlet works, etc. Because of the safety considerations, this type drawdown generally creates the greatest requirement for public information activities.

(6) Aquatic control- During rare cases, a reservoir could be drained to facilitate the elimination of undesirable aquatic life. This type drawdown involves a large coordinated-planned effort between local, State and Federal officials. Normally when a reservoir is drawn down, many secondary problems are solved, such as: maintenance and repair, sediment survey, water quality investigations, etc.

DRAWDOWN SUMMARY- The depletion of reservoir storage, regardless of the reasons, involves the consideration of many factors. To appreciate the difficulty involved in designing reservoir guide curves, it must be understood that the realm of reservoir management is not a straight-forward "cut-and-dried" procedure but a complicated interrelated compromise between many factors. These factors are based on past records of historic conditions which may never occur again. Therefore, flexibility, common sense and a wide area of expertise are essential elements if any success is expected from managing any system of reservoirs for any objective. The chief of reservoir regulation produces the final solution.

MULTIPLE-DISCIPLINED STAFF

The organization, function, and personnel associated with any reservoir regulation staff vary depending upon river basin development, locale, and mission. Therefore, each staff has its own peculiar problems that are associated with particular river basins. Defining these problems and establishing short and long term objectives for their solutions are important factors affecting the staff's function.

A KNOWLEDGE ODYSSEY

The greatest single impact on the function of a water management staff has been the computer. This tool has given the manager the capability to: store voluminous amounts of data; massage the data many ways and many times in a fraction of a second; use sophisticated optimization and system analysis techniques to simulate basin conditions. Currently, (1977) the engineer is in the middle of a technological change or a "retooling" of water management calculation procedures; e.g.- transition from slide rule generated graphs to computer models.

Traditional regulation methods and procedure are simple techniques, by necessity, because of the constraints of manual effort including human error and limited analysis capability. For the most part, these calculations are normally applied only when a problem exists and not on a routine daily basis. This was true because of the time limitation and volume of data handling required to define the quality-quantity of water flowing into and out of Corps reservoirs and the subsequent effects thereof.

The computer capabilities must be responsive to real time use in order to be effective in a daily environment of reservoir regulation. Computer programs or modeling techniques depend upon the hydrologic pulse or reaction time of the depicted reservoir system. This reaction time or hydrologic sensitivity (flood wave celerity, time of concentration of storm runoff, etc.) of the drainage basin will dictate the requirements for staff computer interaction. Programs should be flexible enough so intermediate results can be verified or modified based on new data and/or experience. Every system and subsystem should be scrutinized to determine the most feasible computer modeling techniques that will result into a responsible water management program solution.

SUMMARY
The Corps' multiple disciplined engineers and environmental scientists have been involved in planning, design, construction, and

operation of water resource projects for several decades. However, their capabilities in the ecological, biological and limnological management techniques have just begun to evolve. Water resource management concepts have changed drastically during the last ten years (1967-1977).

Upon completion of a reservoir project, all the ambiguities of the planner and the safety factors of the designer must now be optimized by the water resource manager into an operational plan that: analyzes and develops trade-off functions between water users; balances regional and national goals with local needs; and assures the safety of the surrounding areas and affected populace. Accomplishment of these objectives is the current and future challenges for engineers managing civil works functions.

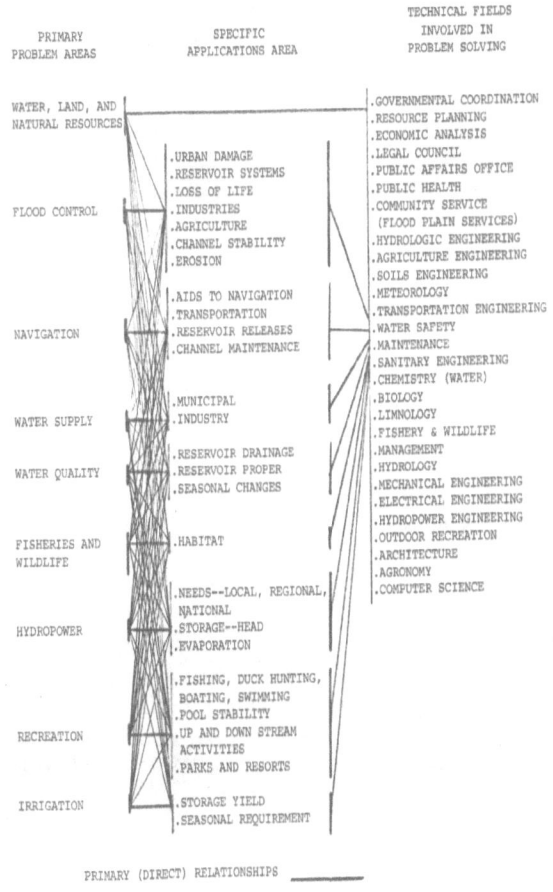

RESERVOIR REGULATION COMPLEXITIES

PRIMARY PROBLEM AREAS	SPECIFIC APPLICATIONS AREA	TECHNICAL FIELDS INVOLVED IN PROBLEM SOLVING

WATER, LAND, AND NATURAL RESOURCES

FLOOD CONTROL
- URBAN DAMAGE
- RESERVOIR SYSTEMS
- LOSS OF LIFE
- INDUSTRIES
- AGRICULTURE
- CHANNEL STABILITY
- EROSION

NAVIGATION
- AIDS TO NAVIGATION
- TRANSPORTATION
- RESERVOIR RELEASES
- CHANNEL MAINTENANCE

WATER SUPPLY
- MUNICIPAL
- INDUSTRY

WATER QUALITY
- RESERVOIR DRAINAGE
- RESERVOIR PROPER
- SEASONAL CHANGES

FISHERIES AND WILDLIFE
- HABITAT

HYDROPOWER
- NEEDS--LOCAL, REGIONAL, NATIONAL
- STORAGE--HEAD
- EVAPORATION

RECREATION
- FISHING, DUCK HUNTING, BOATING, SWIMMING
- POOL STABILITY
- UP AND DOWN STREAM ACTIVITIES
- PARKS AND RESORTS

IRRIGATION
- STORAGE YIELD
- SEASONAL REQUIREMENT

TECHNICAL FIELDS:
- GOVERNMENTAL COORDINATION
- RESOURCE PLANNING
- ECONOMIC ANALYSIS
- LEGAL COUNCIL
- PUBLIC AFFAIRS OFFICE
- PUBLIC HEALTH
- COMMUNITY SERVICE (FLOOD PLAIN SERVICES)
- HYDROLOGIC ENGINEERING
- AGRICULTURE ENGINEERING
- SOILS ENGINEERING
- METEOROLOGY
- TRANSPORTATION ENGINEERING
- WATER SAFETY
- MAINTENANCE
- SANITARY ENGINEERING
- CHEMISTRY (WATER)
- BIOLOGY
- LIMNOLOGY
- FISHERY & WILDLIFE MANAGEMENT
- HYDROLOGY
- MECHANICAL ENGINEERING
- ELECTRICAL ENGINEERING
- HYDROPOWER ENGINEERING
- OUTDOOR RECREATION
- ARCHITECTURE
- AGRONOMY
- COMPUTER SCIENCE

PRIMARY (DIRECT) RELATIONSHIPS _____

SECONDARY RELATIONSHIPS _____

35

CASE STUDIES ON THE FLEXIBILITY OF WATER RESOURCE MANAGEMENT AS RELATED TO RESERVOIRS

Introduction[8]

This chapter demonstrates the necessity for generalized regulation guides as a mandatory requirement for effective, efficient and safe management of a complicated reservoir system.

To manage a water resource system efficiently, the manager is duty bound, by professional ethics, to make the best estimate of current and future situations and react in the best interest of all concerned. This becomes extremely important with systems having multiple purposes, authorized by law, that are readily influenced by seasons, meteorology, politics, multiple agency coordination, etc.

The Kentucky (Tennessee Valley Authority (TVA)) and Barkley (Corps of Engineers) reservoir complex will be used as a vehicle to demonstrate the flexibility required in an actual day-to-day regulation of a flood event.

Authority

Authority for the regulation, by the Department of the Army, of flows from the Tennessee River during flood periods is contained in Section 7 of the Flood Control of December 1944, which states in part: "This section shall not apply to the TVA except that in case of danger from floods on the lower Ohio and Mississippi Rivers the TVA is directed to regulate the releases of waters from the Tennessee River into the Ohio River in accordance with such instructions as may be issued by the War Department.

[8] *Written and presented by this author at the Corps of Engineer's Hydrologic Engineering Center, Sacramento, CA in 1969. After the presentation the River and Harbor Board (33 U.S. Code § 541 - Board of Engineers for Rivers and Harbors; establishment; duties and powers generally) was considering useing the paper for a three hour instruction period in one of their training courses.*

A KNOWLEDGE ODYSSEY

The Secretary of War, on 30 April 1947, formally designated the Division Engineer, Ohio River Division, as the responsible War Department representative.

Pertinent Data

A brief description of the Kentucky-Barkley complex is contained in the following table. In addition to these data, each dam contains a navigation lock that is capable of operation to elevation 375, the maximum flood control pool elevation. The reservoirs are connected by an uncontrolled canal which is used for navigation and to divert water for hydropower generation use.

Table I
Pertinent Data

	Barkley	Kentucky
Drainage area-square miles	17,600	40,200
Uncontrolled area above Dams	7,800	18,800
Reservoir length-miles	118	183
Storage capacity-flat Pool[9]		
Power pool-inches & acre feet	0.62 & 259,000	0.72 & 721,000
Flood control Pool-	3.5 & 1,472,000	
		4.00 & 4,010,800
Conservation-inches & acre feet	1.46 & 610,000	
		1.99 & 1,991,800
Spillway capacity cubic-feet per second	570,000	1,050,000
Spillway gates- type	tainter	fixed-roller lift
Number	12	24 in 3 sections each

[9] Inches of runoff based on uncontrolled area above dam.

37

Size-feet	*55x50*	*40x50*
Power features- units	*4*	*5*

Discharge capacity each-cubic feet per second

Full gate, 42.5 foot head	*56,000*
Full gate, 48 foot head	*52,500*

Operating Objectives

The primary objectives of the flood control regulation by Barkley and Kentucky Reservoirs are:

a. Safeguard the Mississippi River levee system.

b. Reduce the frequency of using the Birds Point-New Madrid floodway[10]

c. Reduce the frequency and magnitude of flooding land outside levees along the lower Ohio and Mississippi Rivers.

The primary control point for operation of the Kentucky-Barkley Reservoirs is Cairo, IL, the Ohio-Mississippi confluence. The magnitude of flooding on the lower Ohio and Mississippi Rivers as related to the Cairo river gage as shown in Figure B of Plate 1. It is obvious that the primary control factor is protecting agriculture interest.

During the occurrence of a flood, the Tennessee and Cumberland Rivers' flood discharge normally precedes the flood discharges from the Ohio and Mississippi Rivers. This is a result of geographical location, in addition to the loss of valley storage[11] on both tributaries because of high degree of control (long reservoir pools). Therefore, the reservoir discharges are normally made just prior to and following the Cairo flood crest. Figure A, Plate 1, is a graphical representation of an ideal flood operation for Cairo.

As shown in Figure A, Plate 1, and in order to release flood waters from the system in advance of a flood, the crest stage should be predicted several days in advance. This stage is referred to as the

---

[10] *The Birds Point-New Madrid Floodway is a flood control component of the Mississippi River and Tributaries Project located on the west bank of the Mississippi River in southeast Missouri just below the confluence of the Ohio and Mississippi Rivers.*

[11] *As the Rivers rise the valley gets wider therefore providing more space for the flood waters to overflow from the river channel.*

"target" stage. During the course of a flood, additional rainfall and/or a more accurate forecast may cause the target stage to be revised. Also, the target stage will be determined by the reservoir releases necessary to limit reservoir storage utilization in proportion to the severity of the flood as indicated by the anticipated Cairo crest and the time of year (growing season, etc.). Normally, during minor and intermediate floods, anticipated crest will include turbine capacity discharges from both projects.

Note: *The computer modeling efforts, heretofore discussed, expanded the accurate (workable) forecast from 2-3 days, that the graphical knowledge manual process provided utilizing two-plus hydraulic engineers to make one forecast per day, to 10-12 days, plus many "what if" iterations per day. As confidence in the model grew with applications the reliable forecast period expanded.*

Another operating consideration is to draw down the reservoirs in advance of a flood. The extent of drawdown will be limited by headwater navigation depths, the natural peak flow for the system and preserving of the total volume under the water surface profile to not less than flat pool volume for both reservoirs at the current guide curve[12] elevation (Plate 2).

The full capacity of the reservoir system will be used if this will avert operation of the Birds Point-New Madrid floodway.

Restraints on Flood Control Operations

Since this discussion is centered on flood control operation, it should be pointed out, however, that this reservoir system is not single-purpose. The relative magnitude of other reservoir functions, in terms of optimum regulation, must be considered. For benefit of this discussion, these purposes, along with other physical limitations, will be considered as restraints to flood control operation.

Some of the possible restraints to flood control operation within the Kentucky Reservoir area are: (1) Flowage easements are modified considerably from June 1st to November 30th; (2) Drainage areas, which are separated from reservoir area by low level dikes, are

---

[12] *Yearly/seasonal pool elevation guide curve to accommodate/coordinate all of the authorized purposes.*

pumped dry every spring. These areas are used for mosquito control and to raise food for wildlife; (3) a minor consideration is the 3-leaf, vertical lift spillway gates at Kentucky Dam. The operation of these gates requires a special crew, other than the regular dam operators, of workers.

Since Barkley Dam was designed and built some 20 years after Kentucky, the design criteria included data from model test (Corps physical Mississippi Basin Model at Vicksburg MS) and from prototype test (Kentucky), therefore, the physical restraints are negligible.

The canal between the two pools has no significant flood control benefit, however, flood waters are diverted from one reservoir the other during the course of a flood. One foot difference between the two reservoirs has been set as the acceptable upper limit to maintain safe navigation. Therefore, if a large in balance of inflow to respective storage should exist, then the canal restrictions could have an effect on flood control operations.

The hydropower function could be a constraint to flood control operations in several ways. Since Barkley-Kentucky hydropower plants are an integral part of the TVA system, full generation and/or full peaking[13] capacity is desired at all times. Therefore, when a flood occurs that requires partial or complete shutdown of releases from the reservoirs, at a time when power demands are high, then a conflict could occur. Power "brown[14]" out enter the operation equation.

Operating Agencies
TVA's River Control Branch, Knoxville, TN
U.S. Weather Bureau's River Forecast Center, Kansas City, MO & River District Office,
Cairo, IL
Corps of Engineers- Mississippi River Commission (MRC), Vicksburg MS; Ohio River

---

[13] Electric power use usually peaks around 7am and 7pm thus additional power is needed for a short time.
[14] Brown out is an intentional or unintentional drop in voltage in an electrical power supply system. Intentional brownouts are use for load reduction in an emergency.

A KNOWLEDGE ODYSSEY

Division (ORD), Cincinnati, OH; Louisville, Nashville, St. Louis, and Memphis Districts.

The Ohio River Division has the responsibility of issuing the notice to begin or terminate the exchange of data. The initiation or termination of the exchange is based on criteria previously agreed upon by the coordinating agencies.

Data Flow to ORD

Plate 3 contains a schematic of the data flow considered necessary for operating the

Kentucky-Barkley system. It's obvious that time is of the essence in the sequential flow of data, since a current analysis of the Kentucky-Barkley-Cairo complex depends on the orderly arrival of data, both observed and forecast, from many sources. Because of the time required to gather, analyze, and transmit data and because of the closed-loop-data flow around Cairo, it is essential that a cursory analysis be made as soon as observed data are available. A more detailed analysis is made as soon as the respective offices transmit their detailed predictions to ORD. Each analysis that ORD makes is made using several assumptions which vary the controllable variables within the system, as well as changing the timing sequence of the uncontrollable events.

Note: *The use of the aforementioned computer model allowed ORD to use its own forecast instead of the National Weather Bureau's river forecast. The Mississippi River Commission (MRC) began using ORD forecast for the Mississippi River below the Ohio River confluence.*

Sample Flood

The following example is not presented for the purpose of evaluating the operation decisions made during the event, but is presented to demonstrate the necessity for having general regulation plans that cover many possibilities and are flexible enough to allow the proper courses of action dictated by a current evaluation of the overall situation germane to the operation. The following general comments pertain to the daily operation of an actual flood event that occurred during the latter part of May and first half of June.

All pertinent actions are recorded in log form.

D-Day -Heavy rains indicate that flood stage will be reached at Cairo, therefore, exchange of data initiated. Situation-it appears that the Cumberland River contribution to total Tennessee-Cumberland is greater than normal. Kentucky and Barkley discharge at turbine capacity. Cairo flood this late in season will likely result in substantial downstream damages. Decisions-it was decided to drawdown both reservoirs in advance of flood using turbine capacity at both plants; plus spillway flow at Barkley which would not raise tail water higher than anticipated from Ohio River backwater.

D+1-Cairo stage 37.5. After an initial analysis of data, the target stage at Cairo was set at 44-45 feet on D+7. The decision was made to increase Barkley spillage and begin spillway discharge at Kentucky. Situation- It would be desirable to increase reservoir drawdown, however, this would cause too large canal flows. In view of anticipated major flooding at Cairo, without restrictions placed on hydropower generation LMVD7 was requested to evaluate benefits for reduction increments and TVA was requested to provide incremented power shutdown losses. Both estimates were received.

D+2-Cairo stage 40.1. Slight discharge reduction was made at Barkley because of tail water control. The Mississippi River at Thebes (upstream of Ohio confluence) appears to be cresting.

D+3-Cairo stage 41.5. Slight additional cut at Barkley for tail water control and decreasing Kentucky to turbine capacity at noon. Situation-Schedule daily cutbacks on both projects for next four days to zero at midnight on D+7 to control crest to approximately 44 feet on D+10 or D+11. Anticipate three or four days of zero flow. De-watered areas Kentucky Reservoir will be flooded.

D+4 -Cairo stage 42.2. Cut backs continuing. Weather activity in Missouri Basin and
subsequently anticipated in upper Mississippi Basin causing some apprehension.

D+5-Cairo stage 42.7. Cut back continuing. Weather immediately west of Ohio Basin is still cause for concern.

D+6- Cairo stage 43.2. Situation- Heavy showers over lower Ohio River and Mississippi River above Cairo. Uncontrolled local runoff *could cause higher Cairo crest. Decision-Since runoff from shower type storm is extremely uncertain, consideration is being given to possible earlier cut to zero. Decision will be made on **D+7**.*

***D+7**-Cairo stage 43.6. Situation-Estimates of runoff from thunderstorm appear high. Current reservoir headwaters within 0.40 feet of forecast. Target stage at Cairo 44 feet on D+9 or D+10. Decision-Decided to cut both projects to zero at noon today. (Normally Barkley cutoff would be sooner because of travel time, but this would cause too much increase in canal flow from Cumberland.)*

***D+8**-Cairo stage 43.7, Situation-Cairo stage steady for 23 hours, however, anticipate .1-.2 foot additional rise. Determined schedule for starting Barkley releases on D+11 and Kentucky releases on D+12. (schedule consists of incremented increases for three days.)*

***D+9**- Cairo stage 43.8. Situation- Pollution problems reported at industry complex below Kentucky Dam on Tennessee River because of zero flow. After coordination with applicable agencies, no change was made in scheduled operations.*

***D+10**-Cairo stage 43.9. Situation- Possible power emergency could occur during afternoon peak load. The load in the Kentucky-Barkley area is being supplied by relatively long lines that are approaching an overload situation, therefore, the afternoon power demands could cause a relay to trip and separate the system, thus causing a blackout in a substantial area. Alternatives-(1) Put Barkley and Kentucky on an emergency standby which would allow immediate full load until connections could be made through an interconnecting link with a neighboring system. This would require from ½ to 2 hours. (This would increase downstream flooding since Cairo is at crest.) (2) Bring one unit on line during early afternoon, instead of midnight. This would not solve the power problem but would lessen probability of separation. Decision-After refining prediction and coordinating with appropriate agencies, ORD decided that one unit could be brought up to no-load speed by 2 pm, could be on-call from 2 pm until 3*
pm and could go on line at 3 pm, without compromise of flood control responsibility. (Barkley was used because of the larger units and because of the longer travel time to Cairo.) Power System remained relatively stable throughout evening.

D+11-*Cairo stage 43.9. Situation-Barkley running one unit and Kentucky discharge zero. Power system came through without incident. Decision-Increase Barkley discharge in steps and Kentucky opening by three hours to avoid rise on Ohio River.*

D+12-*Cairo stage 43.7 and falling.*

Note-*Continued scheduled outflows from both projects (through turbines and spillway} Data exchange terminated on D+ 15.*

Flood summary
During the early stages of the flood, downstream farmers were requesting
through Congressman, that Kentucky and Barkley be cut sooner. However, they were ultimately convinced (?) that the Cairo crest would have been the same.

A complication that could have affected the operation was the flood easement in Kentucky Reservoir. The easement elevation is substantially reduced on I June and the above described flood caused the pool elevation to crest just below the easement.

Summary
The fixed rule criteria evolved from building one project in one basin for one dominant purpose (generally flood control) and continuing the same philosophy as other projects were added. The single projects have grown into systems of multiple-purpose projects that require almost daily reallocation of purposes based on the continuous detailed analysis of large quantities of data. Attitudes and emphasis have changed with time, in that the affected entities have become more aware of their interest, demanding more sophisticated answers to an increasing frequency of questions pertaining to the management of reservoirs. These questions cannot be intelligently answered using firm rule curves for every event, regardless of circumstances.

The computer has made it possible for reservoir system manager to analyze large volumes of data in a very short period. The manager must have the latitude to react based on the "best" solution at the time.

Needs for Future Development of techniques

Technique should be developed so damages and/or benefits can be quickly and accurately evaluated on a compatible base. e.g.-If the probability of a power blackout is so much: what benefit would accrue if such were averted? How can you compare tangible and intangible benefits equably? Should dollars be the sole unit of comparison?

An Engineering Manual on Water Resource Management (not Reservoir Regulation) that would incorporate the thought processes that should be used in evaluating a system that would reflect the present state-of-the-art should be published. The manual should avoid detailed hydrology and hydraulics, be technique, concept and principle oriented, and applicable to dynamic computer utilization as well as flexible and comprehensive regulation rules. The thought portrayed by EM 1110-2-3600, which uses fixed rules for a one reservoir system, should be erased from the Corps and replaced by a more functional document.

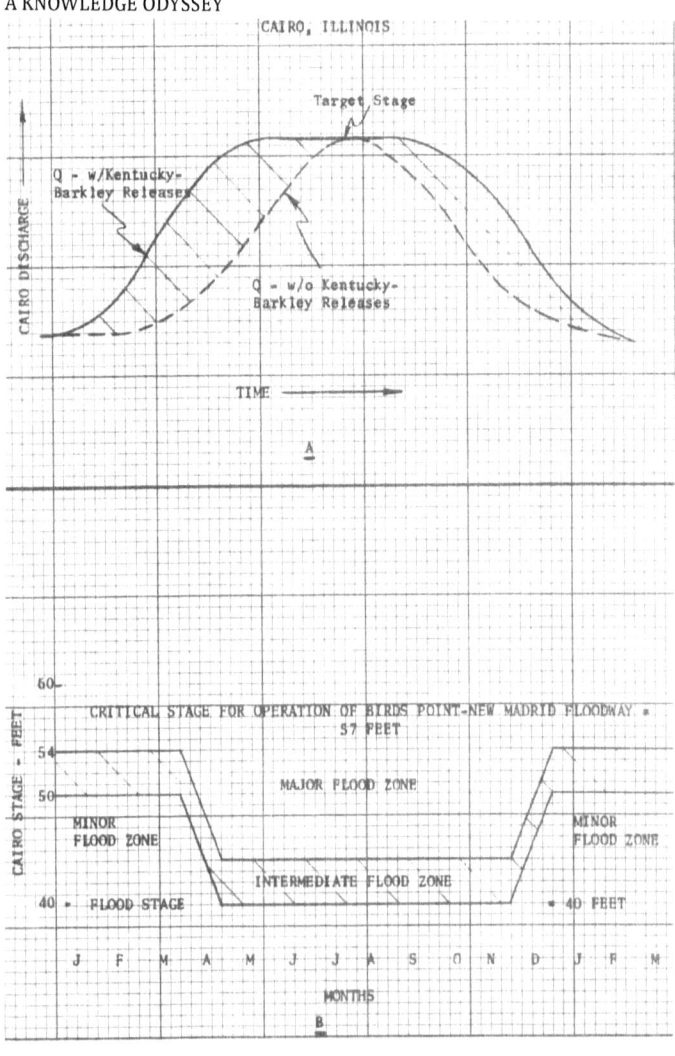

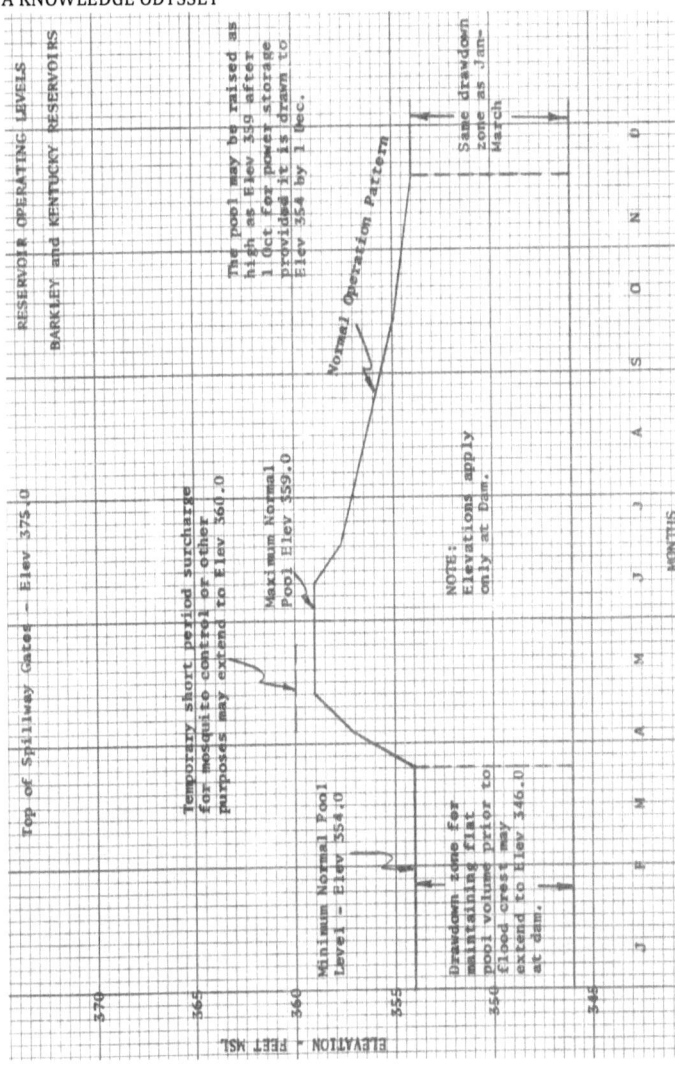

A KNOWLEDGE ODYSSEY

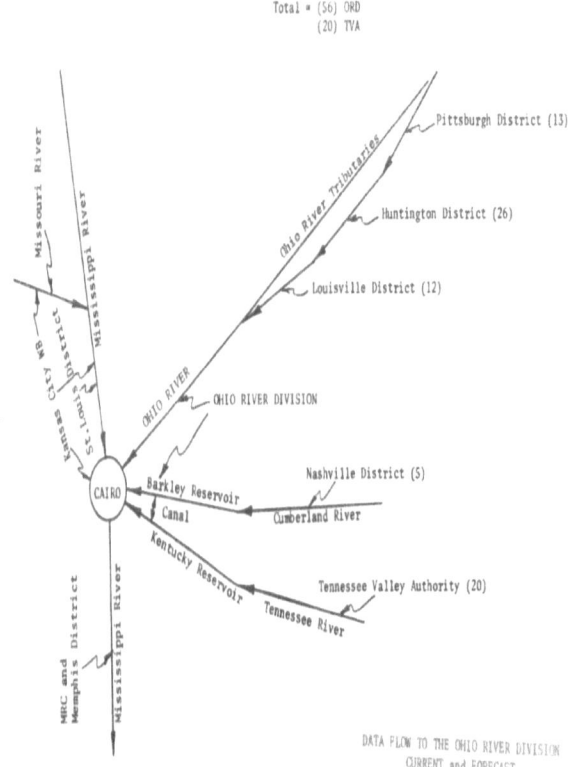

NOTE: (76) - No. of existing flood control reservoirs.

Total = (56) ORD
(20) TVA

Pittsburgh District (13)

Huntington District (26)

Louisville District (12)

OHIO RIVER DIVISION

Nashville District (5)

Barkley Reservoir

Canal

Cumberland River

Tennessee Valley Authority (20)

Tennessee River

CAIRO

Missouri River

Mississippi River

Kansas City WB

St. Louis District

Ohio River Tributaries

OHIO RIVER

Kentucky Reservoir

MRC and Memphis District

Mississippi River

PLATE 3

DATA FLOW TO THE OHIO RIVER DIVISION
CURRENT and FORECAST

48

ICE AND THE OHIO RIVER[15]

The Teays River was a pre-glacial river which drained a large portion of the east central United States. The river met its end when Pre-Illinoisan (Early Pleistocene) ice sheets dammed the region, causing the formation of a large glacial lake, resulting in breached drainage divides and the formation of a new drainage channels. These changes would eventually result in the creation of the Ohio River drainage system, about 12,000 to 13,000 years ago.

Introduction-The Ohio River, which streams westward from Pittsburgh, Pennsylvania, to Cairo, Illinois, is the largest tributary, by volume, of the Mississippi River in the United States. At the confluence, the Ohio is considerably bigger than the Mississippi and, thus, is hydrologically the main stream of the whole river system, including the Allegheny River further upstream. The 981-mile river flows through or along the border of six states, and its drainage basin includes parts of 14 states. Through its largest tributary, the Tennessee River, the basin includes many of the states of the southeastern U.S. It is the source of drinking water for three million people (1976).

The Ohio River is a climatic transition area, as its water runs along the periphery of the humid subtropical and humid continental climate areas. It is inhabited by fauna and flora of both climates. In winter, it regularly freezes over at Pittsburgh but rarely further south toward Cincinnati and Louisville. At Paducah, Kentucky, in the south, near the Ohio's confluence with the Mississippi, it is ice-free year-round. Paducah was founded there because it is the northernmost ice-free reach of the Ohio.

The severe winter[16] of 1976-77 over the eastern United States was related to an exceptionally strong and unusual pattern of winds and pressure anomalies, within the troposphere, persisting from the previous October (1975). The record (ranks no. 1 in Ohio Basin) bitter cold that occurred during January 1977 was additionally due to the southward displacement of the cold polar Low, normally found

---

[15] Author completed course in Ice Engineering at University of Iowa, 1978
[16] Long Range Prediction Group, National Weather Service, NOAA, Washington, DC

over Northern Canada, by what appears to have been the strongest blocking High ever observed over the Arctic region. A record search discovered several cold winters during the 20th century, with the 1917-18 being closest to the number 1- 1976-77 in severity with similar antecedent and concurrent phenomena. The Ohio River was frozen over completely during both of these winters.

Note*: Within the Ohio River basin the winter of 1856-57 is considered to be the worst winter. Historians (Journal-News February 20, 1936) record that this winter was the longest and most severe with Hamilton, OH having zero weather for a period of two months. The Miami River (an Ohio River tributary with confluence just downstream of Cincinnati) was frozen over on November 4th and remained so until the night of February 4th. Official weather records were not kept in the 1850s so this winter does not officially count when ranking the severity of winters.*

Ice on the Ohio River*- The formation, growth and breakup of river ice are a dynamic and complex problem involving the changing state of water, the hydraulics of the river and the associated energy transfer processes. The following presents a general introduction of ice engineering as it relates to the Ohio River.*

To understand ice one first must understand the water molecule. The water molecule can exist in four states: liquid, solid, gas and plasma. The molecule is commonly referred to as: H_2O, yet it can occur in many forms: e.g.-H_2O_2 and many other combinations. Water comes from comets, called "snow balls", from a region near Jupiter called Kuiper. Earth receives many tons of water each year as it flies through the wake of these comets.

Ice forms only after the whole body of water reaches maximum density, which occurs near 4˚Celsus (C) (3.8˚C). At this point the river becomes more or less isothermal with depth, regardless of flow velocity. This means that as the water surface is further cooled toward the 0˚ freezing point the lighter water on top is the coldest and, therefore will remain on the surface, in the absence of turbulence. An important point to remember is that the phenomenon of freezing and thawing of ice takes place in an extremely narrow range of temperatures (-0.1˚C to + 0.1˚ C). The water-ice interface is always at 0.00°C. Studies have shown that pollution has very little effect on these temperatures values in large bodies of water.

There are two basic forms of ice that occur on the Ohio River: (1) lake ice; and (2) frazil ice (slush). The two ice forms are completely different in terms of formation processes, crystalline structure and physical characteristics.

Lake ice forms in place. The ice crystals are long and slender, regimented always in a vertical plane. After the river has cooled to 4°C, the quicker an ice cover can form, normally the thinner it will ultimately be. The reason for this is that with the surface temperature at 0°C (-) and the remaining water from 0° C to +4°C (-), an enormous amount of thermal energy is entrapped in a body of water like the Ohio River. The term "warm water" means the water temperature is at least 0.2° C. This important for two reasons: (1) the ice cover will insulate the water from colder air temperatures and therefore, trap the heat energy in the water as well as suppress the rate of further ice formation (Ice is formed at the ice-water interface only), and (2) the ice will decay faster because the melting is almost entirely caused by the heat energy in the water and not the air.

Lake ice is relatively easy to break or shear because it generally fractures along the crystalline boundaries which are oriented vertically with the crystal length nearly equal to the ice thickness (undisturbed ice). The crystals will not cling to a vessel or dam because the "open" water supplies enough energy to melt the ice.

During the winter of 1976-77, Ohio River ice cover was of the lake type. The energy stored in the water prevented ice formation to occur below the navigation structures by simply melting the ice. Turbulence from the dams produced a continued supply of warmer water to the surface.

Frazil ice (slush ice) forms at the surface in turbulent water that is cooled to -0.01° C to -0.03° C (super cooled) temporarily when coming in contact with subzero air. This occurs when a body of water has normally lost most of its heat energy and is therefore at or near 0°C. Turbulence must be sufficient to carry the ice crystal away from the surface once it's formed. Frazil ice crystals are very small and flat, ranging from 5 mm to 15 mm diameter (smaller than a dime).   The crystals normally cling to everything (one another, vessels, river bottom, submerged banks, structures, water intakes, etc.)

Floes[17] are formed further downstream below the dams or at the source of turbulence as the crystals attach to one another. Ice cover becomes an accumulation of ice floes which create a very rough and tough ice barrier.   Frazil ice creates the worse condition for navigation because it's long lasting (river heat energy almost dissipated, therefore no melting potential), it clings to all surfaces, and it forms "hanging" dams (Markland-winter 1978).

Ice breakup and movement- *Generally, the ideal situation to eliminate an ice cover is for the ice to decay (rot) in place. However, this does not normally occur on large rivers such as the Ohio. Ice breakup and movement can result from several physical events, namely: (1) a small increase in flow from rain and/or snowmelt runoff; (2) an increase in average air temperatures accompanied by mechanical breakup from boats; and (3) a combination of the above. Once ice begins to move it is best to keep it moving, otherwise a jam will obviously occur. Freedom of ice movement from one pool to another should be controlled insuring that the lowermost pool is always unobstructed.*

Ice jams are likely to occur where:

(1) The slope of the river changes abruptly from steep to gentle;

(2) The point of confluence of two rivers;

(3) The river width diverges as opposed to converges;

(4) The river curves greater than 110-115 degrees;

(5) The tributary meets the Ohio backwater profile;

(6) The ice floes are from 1/3 to ½ the size of the distance between bridge or dam gate piers.

---

[17] *An **ice floe** (ice float) is a large pack of floating ice. Such ice floes are found in:*

- *Drift ice, any type of sea ice not attached to land*
- *Ice dam (ice jam), a blockage of ice in a river starting with an ice floe in the river*
- *Ice stream, a type of fast moving glacier*

Ice flow, the movement of ice, is a misspelling of ice floe.

Areas of the river where islands are formed are also susceptible to ice jams because the same hydraulic forces that cause sediment deposition also cause ice deposition. Sediment falls out on the inside of a curve while ice is initially deposited on the outside at the surface profile (one is inverse to the other).

Destruction of ice jams- *Once a major ice jam is formed and settled, it is extremely difficult to destroy by mechanical or other means. The best chance of success exists only in the early stages of formation. The destruction of an ice jam by mechanical means (boat, etc) should proceed on the downstream edge at the point of highest stress concentration. Destruction by explosives (if feasible) should begin at the anchoring points such as islands, sandbars, etc. Pools downstream of the dam should be capable of handling the released ice before starting to destroy the jam. Sometimes the quest to resume navigation at a particular project overshadows this very important point.*

Ice suppression devices- *Bubbler systems work by inducing warm water against an ice cover by means of a bubbler-driven water jet, thereby melting or suppressing the growth of the ice cover. Successful operation of an air bubbler system requires a continued supply of warm water. (Warm water is defined as water with a temperature of at least 0.2°C or above.)*

There are two conditions that can normally exist to cause the bubbler system to be ineffective: (1) if the average flow velocity exceeds 1.5 to 2.0 feet per second, the heat transfer potential at the water ice interface is as fast as can be induced by the bubbler system, and (2) if the water body temperature is 0.1°C or below, there is simply very little available heat energy, therefore, no melting potential. (Studies have shown that the temperature of the air source for a bubbler system has a negligible effect on the system's operational efficiency).

Surface effect (air cushion) vehicles (SEV) appear to be most promising (1979 era) in the field of lake ice breaking. Tests have shown that a SEV can easily break a channel with less resistance than a conventional icebreaker. Experiments and analytical studies have shown that ice breaking with SEV's occur in two ways: (1) at low speeds the ice fails because the cushion pressure depresses the water under the ice sheet leaving large sections of ice unsupported, therefore, the ice breaks under its own weight; (ice is only stable in compression forces) and (2) at high speeds the SEV creates a standing wave (like any vessel) which moves at the speed of the craft and therefore, continually breaks the ice at the wave's crest.

An SEV can progress continually through an ice field (lake ice) breaking it as thoroughly as an icebreaker ship if: (1) the SEV's cushion pressure, expressed as head, slightly exceeds the ice thickness; and, (2) the weight of the SEV exceeds that weight which will create ultimate collapse of the ice cover.

FLOOD FREQUENCIES MODIFIED BY RESERVOIRS

Accordingly, the poet should prefer probable impossibilities to improbable possibilities. Aristotle

Introduction This is a tutorial presentation about frequencies' applications in water resource management. No theories, formulas, equations or curves, just a discussion of the basics and how the end results are used as tools for design, forecasting natural events[18], calculating project benefits after the event and, defining the "ballpark" within which to operate projects based upon analysis of historic data.

Nomenclature Possibly you can remember when a really big rain, be it from a hurricane or a large frontal system or thunderstorm, hit your town. If flood conditions occurred because of this rain, then you might have heard the radio or TV weatherman say something like, "This storm has resulted in a 100-year flood on the Guyandotte River which crested at a stage of 20 feet." Obviously, this means that the river reached a peak stage (height) that happens only once every 100 years, right? A hydrologic engineer or water resource engineer would answer, "Well, not exactly." Engineers don't like to hear a term like "100-year flood" because, scientifically, it is a misinterpretation of terminology that leads to a misconception of what a 100-year flood really is.

Instead of the term "100-year flood", a hydrologic engineer or water resource engineer would rather describe this extreme hydrologic event as a flood having a 100-year recurrence interval. What this means is described in detail below, but a short explanation is that, according to historical data about rainfall and stream stage, the probability of the Guyandotte River reaching a stage of 20 feet is once in 100 years. In other words, a flood of that magnitude has a 1 percent chance of happening in any given year.

---

[18] *Floods, droughts, seasonal or routine reservoir regulations.*

What is a recurrence interval Most frequency methods desire the ideal 100+ years of observed data to analyze but there are methods, called regression equations[19], which can be used to estimate frequencies for lesser years. Thus the hydrologic engineer or water resource engineer will have more confidence (confidence level can be calculated) on an analysis of a river with 30 years of record than one based on 10 years of record.

Recurrence intervals for the annual peak to a given location change if there are significant changes in the flow patterns at that location, possibly caused by an impoundment or diversion of flow. The effect of an upstream reservoir project or system on peak flows is generally much greater for high recurrence interval floods, such as 25- 50- or 100-year floods. The reservoir system or systems are designed to store water until the flood wave has passed the downstream applicable control stations.

A Frequency method This chapter presents the basic procedures used by the Corps of Engineers to develop frequency curves.

Flow frequency procedures are based on methods developed by astronomers, surveyors and others during the 18th and 19th centuries. These early scientists developed statistical methods by which data variations could be studied in the analysis of measurement errors. However, the statistical methods were not effectively applied to large volumes of raw data, such as hydrologic data, until after 1914, the year an American engineer, Allen Hazen, invented probability paper[20]. With probability paper, data extremes could be analyzed relatively easy and alternatives could be studied without extensive computation.

Flow frequency computations based on statistical analysis of past records allow the engineer to make better estimates of future flow conditions than is indicated by the raw data. There are many

---

[19] *Use up to 12 or 13 parameters-e.g.-basin characteristic, stream slope, rain fall, type of soil, similarity with a recorded basin, etc.*
[20] *Graph paper with vertical and horizontal rules, the latter spaced evenly and the former according to a scale that allows a plotted probability curve — usually that of a normal distribution — to appear linear. (straight).*

frequency determination methods. Common methods give acceptable uniform results within the range of data. However, if adequate data are not available, or if frequency curve extensions are made beyond the range of data, there can be appreciable differences in results from the various methods.

Also, different results can occur when the broad engineering problems are overlooked and the study is concentrated on the statistical tool as an end in itself.

Flood frequency determination is one of the technical methods that have experienced separate agency development over the years. Each agency developed its methods based on their respective needs. The Water Resource Council's[21] (WRC) Hydrology Committee made an extensive evaluation study of different frequency methods in an attempt to adopt a uniform technique. Their recommendations are contained in a WRC Publication, "A Uniform Technique for Determining Flood Flow Frequencies", dated December 1967.

Summary If time and money were available, a more sophisticated approach could be taken. Synthetic periods of record (each 100 years in length) could be generated using stochastic hydrologic methods[22]. E.g. - Ten 100 year periods could be generated. Frequency analysis methods would be applied to each of the 10 periods separately obtaining ten different frequency curves. The computed average from these curves would be the adopted frequency curve for natural conditions. The curve modified by reservoirs could be generated by the same stochastic methods.

Sophisticated methods sometimes cause the engineer to become "hung up" on the statistics rather than the broad engineering considerations behind the use of frequency curves.

---

[21] *U.S. Code› Title 42 › Chapter 19B › Subchapter 1*
[22] *The level of knowledge to use these methods is at the master degree level in hydrology and geosciences. Advanced statistical and stochastic methods are commonly used in analyzing hydrological data series and stochastic modeling of hydrological processes at spatial and temporal domain using data from hydrology and geosciences.*

Basic frequency curve calculations must be done using data with the same base conditions and the data record should be continuous (unbroken chronological record) for the study period. Frequency curves can be "adjusted" by historical data, regional correlation (preferably the same drainage basin) and correlation with data from longer period-of-record stations from similar hydrologic drainage basins. This is done with regression equations which are beyond the scope of this presentation.

There are other considerations such as biased and/or skewed data, e.g. - drought vs wet periods, etc.

Frequency curve accuracy as a representative sample of the hydrologic characteristics of the basin being studied can only be measured or expressed in terms of the validity of the basic data and statistical method used in the data analysis.

CARBON DIOXIDE (CO₂)

"Both shale gas and conventional natural gas have a larger greenhouse gas footprint than do coal or oil, especially for the primary use of residential and commercial heating". - Energy Science & Engineering July 21, 2014

Introduction- A study[23] published in the journal *Science* found that government biofuel policies rely on reduction in food consumption to generate greenhouse gas savings. Decreasing the amount of food that people and critters eat produces less CO_2 they breathe out or excrete as waste. The reduction in food that's available for consumption, rather than any inherent fuel consumption drives the decline in CO_2 emissions in government models, the researchers found. Crops diverted from food to biofuels are not replaced by crops elsewhere. E.g. - 50% of the net calories diverted to make ethanol are not replaced through the planting additional crops. Both models used by U.S. Environmental Protection Agency and the California Air Resources Board indicate that ethanol made from corn and wheat generates modestly fewer emissions than gasoline. These facts of lower emissions are buried in the methodology and are not stated.

In the European Commission's model even greater emissions in the reduction in both the quality and quantity of food with food of lesser nutritional value causes ethanol making with wheat to be 46% higher than gasoline and corn ethanol 68% higher emissions.

The Source of Earth's CO_2- Anyone can observe the vapors from dry ice (CO_2), at room temperature, as they fill up a container and fall over the edge like a water fall. Since CO_2 (390 ppm[24]) resides in the upper atmosphere, many researchers have searched for the mechanism which could lift CO_2, a gas 2.44 times heavier than the atmospheric gases, to the upper reaches of the atmosphere. One solution comes from the observation of data collected from comets.

---

[23] *Searchinger, T., Edwards, R., Mulligan, D., Heimlich, R., Plevin, R., "Do Biofuel Policies Seek to Cut Emissions by Cutting Food", Science, 2015*
[24] *Number of parts per million parts*

The water and carbon dioxide molecules from comets that originate near the planet Jupiter (the Kuiper Belt) has made Earth an oasis for fauna and flora. Ground truth data came on November 4, 2010, the date of the EPOXI and Deep Impact mission flyby of the comet Hartley 2[25], as CO_2 was observed escaping, taking with it "snowballs[26]" from the comet.   University of Maryland Astronomer Michael A'Hearn, the principal investigator, said: "When warmed by the sun, dry ice [frozen carbon dioxide] deep in the comet's body turns to gas jetting off the comet and dragging water ice with it." These finding supports a controversial, but accepted idea, that comet impacts for billions of years provided most of the water in Earth's oceans. "The smaller comets from the Jupiter's region (Kuiper Belt) impacted Earth relatively gently, shattering high in the atmosphere and delivering most of their organic molecules intact"[27]. In retrospect the "snow ball" comet[28] brought water, CO_2 and microbes to an early Earth. This process continues today.

The comet 67p/Churyumov-Gerasimenko[29] (67p) has been studied[30] in detail using data gathered from the Rosetta and Philae spacecraft since September 2014. The comet's body has an irregular "duck shape" with distinct features such as boulders, craters and steep cliffs that are clearly visible. Dr Wallis, and colleague Professor

---

[25] Hartley 2 is a comet from the Jupiter region referred to as "Snowballs". These comets have the identical water molecule that's found on Earth. (NASA)

[26] Another type comet called "Icebergs" is from the Oort region located on the outer edges of the Solar System. These comets have a molecule different than the water molecule found on Earth.

[27] The Science and Technology Directorate at NASA's Marshall Space Flight Center, May 18, 2001

[28] Carnegie Institution (2012, July 12). Solar system ice: Source of Earth's water. Science Daily. Retrieved July 13, 2012, from sciencedaily.com/releases/2012/07/120712144743.htm?

[29] Rosetta mission-NASA/Jet Propulsion Laboratory, ScienceDailey, 22 January 2015

[30] European Space Agency-Royal Astronomical Society, "Do micro-organisms explain features on comets", ScienceDaily, 5 July 2015

Wickramasinghe, Director of the Buckingham Centre for Astrobiology, conclude that these features are all consistent with an ice mix with organic material. As the comet approaches the Sun the temperatures are rising and the micro-organisms will become more active. Rosetta has already shown that the comet is not seen as a deep freeze object and as it approaches the Sun it will become more hospitable to micro-life than our Arctic and Antarctic regions. The researchers cite further evidence for life in the detection of abundant complex organic molecules on the comet's surface by Philae and in infrared images taken by Rosetta.

The comet, 67p continues to spew out water and CO_2 with peaks of each appearing to be based on night and day or seasonal (summer-winter) effect. There is a clear distinction between the "summer" and the "winter" (hot and/or cold local conditions) atmosphere. The "summer" zone is dominated by water ejection and the "winter" zone, which is about one kilometer away by CO_2 ejection.

Research shows that the breakup of another "snowball[31]" comet, Comet LINEAR[32], was likely made up of water with the same isotopic composition as water found here on Earth. The finding supports a controversial idea that comet impacts billions of years ago could have provided most of the water in Earth's oceans.

Space is populated by the water and CO_2 molecule by comets from the Kuiper Region and Earth will fly through these many contrails absorbing these molecules.

In retrospect the "snow ball" comet brought water, CO_2 and microbes to an early Earth and this process continues today.

---

[31] *Comets from deep space formed in an extremely cold region thus are called "ice bergs" while comets formed in the Jupiter region, a more moderate cold, are referred to as "snowballs".*
[32] *August 2000*

CO_2 Importance to Earth- The reality of how essential carbon dioxide (CO_2) is to the survival of the fauna and flora is being over shadowed by the fears of a few who want to "kill the goose that laid the golden egg".   To comprehend the importance of the CO_2 component to the fauna and flora coupled with the miniscule effect that it has upon the whole global thermal budget, one must understand the constituents of the global climate's workings.

It appears that the modification or throttling CO_2 discharges from facilities that process fossil fuels will have no measurable influence upon the thermal health of the globe when compared to the importance of water vapor and dust, et cetera.   The latter two components form a thermal blanket in the outer reaches of Earth's atmosphere. In fact Earth's thermal history suggest that an increase in CO_2 within the influence of the Hydrologic Cycle (aka Water Cycle[33]) will help the fauna and flora to thrive and result in no significant influence upon Earth's overall thermal budget.

Regional climates are cyclic, e.g. Σ 6°C or +/- 3°C[34] fluctuating every few decades. It appears that checks and balances have controlled this scale of variation since about two billion years ago.

Wetlands absorb carbon from CO_2 at a higher rate when CO_2 is more plentiful[35]. Thus the more CO_2 available the more the flora eats! As a matter of fact trees use water more efficiently as the CO_2 rises in

---

[33] *The hydrologic cycle (water cycle) describes the existence and movement of water on, in, and above the Earth. Water on Earth moves continually through the hydrologic cycle of evaporation and transpiration (evapotranspiration), condensation, precipitation, and runoff. Its final destination is usually the oceans. Evaporation and transpiration contribute to the precipitation over land. The water cycle has been working for billions of years and life on Earth depends upon it continuing to work. (US Geologic Survey)*
[34] *Celsius scale*
[35] *Atomic weight of CO_2 molecule is 44 compared to water (H_2O) 18 and the atmosphere 14+ (nitrogen is 14.01)*

the atmosphere. This is caused by the CO_2 intake "vents" on the leaves contracting in response to the increased CO_2.

One of the accepted basis for CO_2 concentrations in the atmosphere is presented as the "Atmospheric Carbon Dioxide" graph (Keeling Curve) depicting continuous measurements observed at Mauna Loa, HI, at an elevation 3,397 meters (11, 141 feet) above mean sea level (msl) since 1958. Research has found no CO_2 observations recorded above this altitude[36] or at the equator. The assumption to date is that the rise in the observed CO_2 values has been attributed to a sole source, entropy from the use of fossil fuels. The possibility that a part of the measurements were from the heavier than air CO_2 molecule deposited by comets over time settling through the lighter atmosphere to the Water Cycle region was not addressed as to the source.

Charles Keeling alerted the world that a concentration of CO_2 in the atmosphere was increasing. His graph showed not only the annual increase in CO_2 levels but a season-by-season oscillation in the concentration.

During the summers in the Northern Hemisphere the Earth's flora breathes in CO_2 and retains the gas. In the winter the gas is exhumed as the summer's leaves decay. Since green leaves were the facilitor and CO_2 was rising, the Earth must be getting greener.

In the 1980 forest biologists reported an increase in growth rates of trees and forest densities, particularly the Douglas fir in British Columbia, Scott pines in Finland, bristle pines in Colorado and the tropical rain forest around the globe.

NASA scientist Compton Tucker mapped global vegetation changes by calculating "Normalized Difference Vegetation Index" (NDVI) from data gathered by satellite. This data confirmed Keeling's data. The globe is becoming greener. Satellite observations confirmed that the entire globe was getting greener.

---

[36] NASA launched an orbiting Carbon Observatory-2 (July 2, 2014) to study CO_2 in Earth's atmosphere. A two-year mission to unravel key "mysteries" about CO_2.

Using NDVI it was determined that Earth's ecosystems were acting as net carbon sinks. Since they were absorbing more carbon than emitting the effect has slowed the CO_2 increase in recent years. Satellite data has confirmed that greening has been occurring for the last three decades. Between 1982 and 2011, 20.5% of the global flora got greener, while 3% grew browner and the remainder showed no change.

An increase in CO_2 will accelerate growth in flora, even without a warmer climate since most floras has trouble capturing CO_2 from the atmosphere because of its scarcity. The end result is that the flora tolerates heat better when CO_2 levels are higher.

Homo sapiens use of fossil fuel has caused the greening of the Earth in three ways: (1) not using firewood as fuel; (2) warming regional climates; (3) raising the level of CO_2 which has increased flora growth rates.

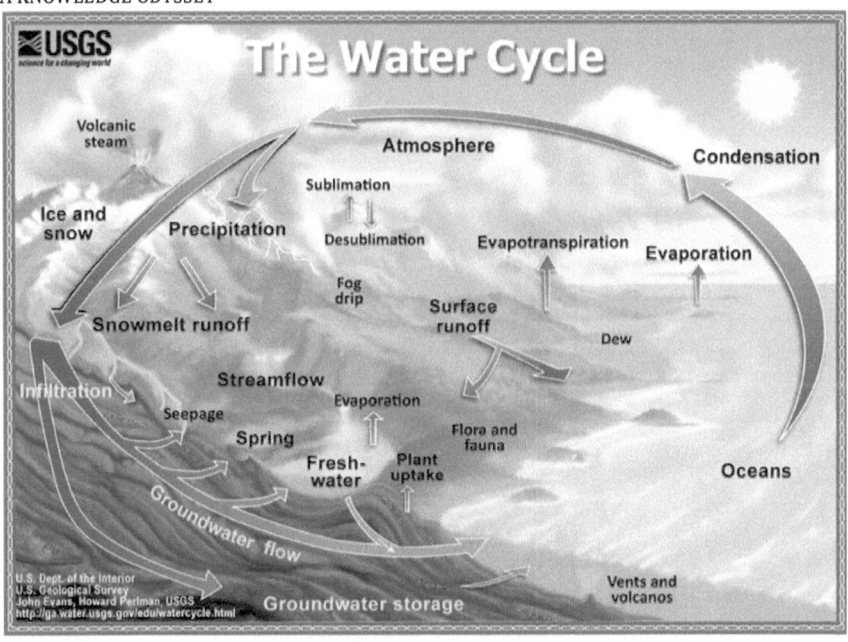

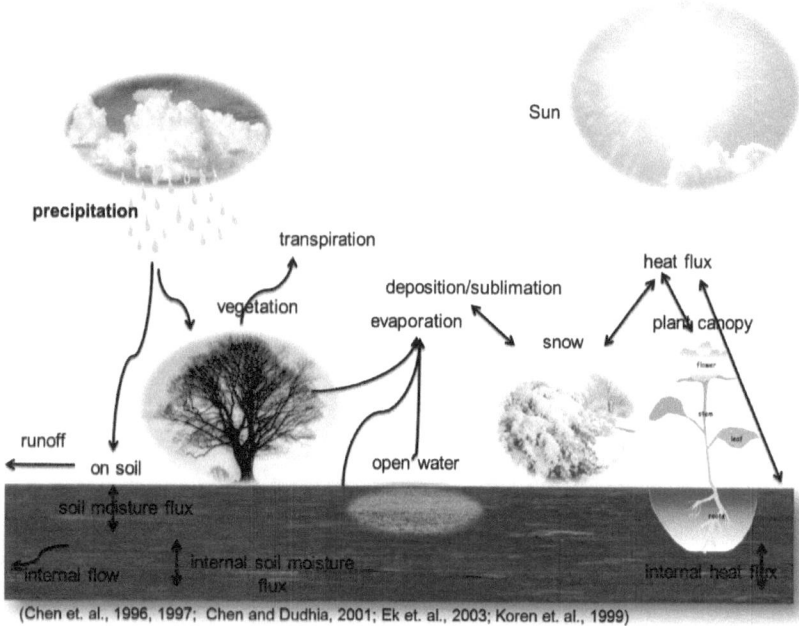

(Chen et. al., 1996, 1997; Chen and Dudhia, 2001; Ek et. al., 2003; Koren et. al., 1999)

It appears that **all** the breathable oxygen (O_2) and all of Earth's carbon came from "Snowball" comets in the form of CO_2. Because the CO_2 molecule is heavier[37] than the atmospheric gases, the CO_2 will slowly sink toward the surface while meandering toward the equator because of gravity and gyroscopic forces caused by Earth's

---

[37] *Carbon dioxide weighs 44 atomic mass units (amu). The atmospheric column of 78.09% nitrogen (14.01 amu); 20.95% oxygen of 32 (2*16 amu); 0.93% argon of 36 (18* 2 amu); .039% CO_2; and 1% water vapor; = 25.3 amu.*

rotation. As a result the flora at the equator becomes more robust, producing an increase amount of O_2 thus making the fauna more robust.

An example of this phenomenon occurred from about 200 million years ago (Triassic period) to about 65 million years (end of the Jurassic period) ago when the dinosaurs roamed the Earth. Also Nature used this process to produce most of the fossil fuels that we harvest today.

This one observation of the comet Hartley 2[38] has begun to solve one of the most studied mysteries of atmospheric science. It's understandable how a heavier gas can be propelled (turbulent diffusion) to the upper influences of the Hydrologic Cycle (approximately one-hundred-thousand feet altitude at the equator) because of the mixing forces (thermal, wind and orographics effects) within the water cycle. That is how a gas molecule such as carbon dioxide (CO_2), which is approximately two point four (2.44) times heavier than the air column, at Earth's surface, can defy gravity and rise to the upper boundaries of the Water and Carbon Cycles[39]. There is no evidence which suggest that known natural mechanisms that can lift the CO_2 generated on Earth's surface higher than the boundaries of the "water cycle". In other words the CO_2 molecule **does not defy gravity**!

Here is an example! Lake Nyos, located in Cameroon, Africa is on a list of the 10 most dangerous lakes on Earth. Yellowstone Lake in the USA is also on that list. These lakes are positioned over large

---

[38] *August 2000-See photo front cover.*
[39] *The global carbon cycle is usually divided into the following major reservoirs of carbon interconnected by pathways of exchange: atmosphere, terrestrial biosphere, oceans (including dissolved inorganic carbon and living and non-living marine biota), sediments (including fossil fuels, fresh water systems and non-living organic material, such as soil carbon) and the Earth's interior, carbon from the Earth's mantle and crust. These carbon stores interact with the components through geological processes.*

reservoirs of magma that releases carbon dioxide (CO_2) into the water (H_2O). The CO_2 reacts with H_2O creating carbonic acid (H_2CO_3) a lethal gas to all living critters that require oxygen for survival.

On the night of August 21, 1986 something, probably an earth slide, in Lake Nyos forced the CO_2 to the lake surface, filling up the lake basin and overflowing, following the nap-of-the-earth into the adjacent villages asphyxiating 1,700 people and 3,500 animals.

Yellowstone Lake, USA, a crater lake located over a very active volcanic region thus the geysers, has released CO_2 that suffocated large animals that have bedded down in low lying areas for the night.

Eyjafjallajökull volcano in Iceland erupted in 2010 and released more CO_2 into the atmosphere in **one (1) second than all the fossil fuel driven devices have in the last 180 years**. This will put some magnitude toward Homo sapiens's activities. Iceland alone has about 45 active volcanoes. There are a few hundred more on Earth.

This volcano was puny when compared to the eruption of Tambora 200 years ago. Tambora's ejection of sulfurous aerosols into the atmosphere created a year without a summer in Europe and the eastern North America in 1816. Recently, scientists[40] have proposed that the eruption of Toba on the island of Sumatra 74,000 years ago was the most destructive eruption ever recorded. Historical records of volcano eruptions are very incomplete.

Vast ranges of volcanoes hidden under oceans are the gentle giants of Earth as they dwell along the mid-ocean ridges. They flare up in regular cycles, from two weeks to 100,000 years, apparently triggered by the short and long term changes in Earth's orbit. Eruptions are exclusively during the first six months of each year, apparently caused by Earth's solar orbit and tilt and direction of its axis (seasons). There is no evidence that ocean volcanoes emit large amounts of CO_2.

---

[40] Self, Stephen, Gertisser, Ralf, "Tying Down Eruption Risk", Nature Geoscience, 2015

Note: The Tambora eruption had one famous result: Had it not been for the cold and wet weather throughout Europe Mary and Percy Shelly and Lord Byron might not have spent the summer of 1816 telling ghost stories around a log fire in Lake Geneva, and Mary Shelly might never had turned the best of these stories into a famous book, "Frankenstein".

The referenced researchers suggest that 90 % of the volcano risk worldwide is in five nations: Indonesia; Philippines; Japan; Mexico; and Ethiopia.

Mainland North America has Crater Lake, created by a Tambora size eruption 7,700 years ago, while Yellowstone National Park was ground zero for a major eruption 640,000 years ago. This eruption blanketed much of North America with ash. Long Valley caldera east of California's Sierra Nevada, within the town of Mammoth, is considered an active super volcano though its last activity was 760,000 years ago.

Smaller volcanoes, Mounts Rainier and Hood, in Washington and Oregon, respectively, are considered active, while California's Mount Lassen erupted just 100 years ago.

Besides determining the source of Earth's CO_2, an understanding of the forces that the atmospheric gases are subjected to within the air column is explained in detail. The "why" question trumps the "how" question in this discussion.

Within the five principal layers of the atmosphere determined by thermal characteristics are several layers determined by other properties. For instance the planetary boundary layer is part of the troposphere that is nearest the Earth's surface and is directly affected by turbulent diffusion. During the day it is usually well mixed, and at night it becomes stably stratified with weak mixing. The depth of the planetary boundary layer ranges from about 100 meters on a clear calm night to 3,000 meters or more during the afternoon in dry regions. This altitude approximates the tree line elevation.

The upper altitude limits for the Water and Carbon Cycles is the Stratosphere (Ozone Layer), which extends to about 35 kilometers (km) (110,000 feet) altitude. This is the upper limit for any CO_2 being propelled from Earth's surface by a thermal event such as a thunder storm (cumulonimbus cloud). Between the Water Cycle and the Homosphere there are no other lifting mechanisms.

The next mixing layer defines the homosphere (includes the troposphere, stratosphere and mesosphere) and heterosphere layers. Here gases are mixed by turbulence. In the mesosphere at about 100 km, the gases composition varies with altitude because the distance that particles can move without colliding with one another is large compared to the size of motions that cause mixing. This allows the gases to stratify by molecular weight with the heavier ones such as oxygen and nitrogen only occurring near the bottom of the heterosphere. The upper portion is almost completely composed of the lightest element, hydrogen.

The only time in Earth's history that global temperatures increased, beyond the limits of normal variations, was a result of the entire globe being covered with ice. CO_2 from the "Snowball" comets became trapped in the atmosphere above the ice layer. Scientists believe that this event was about one-half to three-quarters of a billion years ago. This thermal "blanket" eventually caused Earth's surface temperature to raise melting most of the ice. Except for this one occurrence, the **default** condition for Earth has always been **ice.**

Recently an international team that includes scientists from Johannes Gutenberg University Mainz (JGU) has published[41] a reconstruction of the climate in northern Europe over the last 2,000 years, back to 138 BC, based on tree ring density measurements from sub-fossil pine trees originating from the Finnish Lapland. This region's climate is moderated by the Atlantic Meridional Overturning

---

[41] *http://www.sciencedaily.com/release/2012/07/120709092606.htm? ; published in the journal "Nature Climate Change".*

Circulation (Gulf Stream). The "team's" interpretation of the data suggest that this regional climate for the "past two millennia has been towards climatic cooling" of minus 0.3 degrees centigrade (°C). An examination of the reconstructed temperature data suggests that the cold and warm phases of this regional climate can vary as much as plus or minus 3° C, from the mean. This means that a total of 6°C, from maximum to minimum, can be expected from warm periods to mini-ice ages. The bottom line is that an ice age or warm period can only be identified by current interpretation of scientific data when each is well into their respective cycles.

There is no evidence that the "Brownian Motion[42]" has any significant influence in the definition of the movement of the CO_2 molecule within the atmospheric gas column. The atmospheric boundary of the Water Cycle is not a closed homogenous fluid as the Brownian experiment environment was conducted, but a fluid column that changes density, content and thermal mixing affects with altitude.

During a lecture on wine attended by this author in the California wine country, in 2010, the lecturer described an experiment conducted by the makers of champagne suggested that the best way to store an opened bottle was in the refrigerator with the cork missing. The CO^2 will not escape.

Scientists do not have the thermal evidence or scientific knowledge to determine if Earth's climates are continuing to recover from the last ice age (the 4th cycle 12,000+ years ago) or beginning a fifth. The aforementioned research project suggested that one region of

---

[42] *Brownian motion is the presumably random drifting of particles suspended in a fluid (a liquid or a gas) or the mathematical model used to describe such random movements. In both cases, it is often mathematical convenience rather than the accuracy of the models that motivates their use. This is because Brownian motion, whose time derivative is everywhere infinite, is an idealized approximation to pretty stale place to live without its actual random physical processes, which always have a finite time scale. MÅorters, Peter; Yuval Peres (25 May 2008).*

the globe is cooling. (How this regional thermal budget is managed will be discussed later.) However, a millennium of climate data[43] may only make one definable point on the historical climatic summation curve? Since the "tree study" suggest that climatic data for every year is different from every other year in every region of Earth, predicting future climatic trend is more art (reaction) than science, since there is NO scientific way to calibrate the models that are used to make these forecast[44].

According to the NOAA, scientific measurements of the CO_2 in ice cores, or cylinders of ice, showed that the CO_2 level prior to the Industrial Revolution was 278 parts per million. In addition, the levels did not change more than 7 ppm between 1000 and 1800 A.D.

In the early 1600's when Europeans first arrived upon these shores, the Appalachian Mountain range was covered with mature trees (virgin forest). The Appalachians run from Maine to Georgia. Most of the "virgin" timber has been harvested while some second growth forest has been preserved as National Forest. Hardwood trees in these forests will take from 150 to 500 years to develop old-growth characteristics in one or two generations of trees.

Cathedral State Park in West Virginia consists of only 133 acres of over 170 species of vascular flora that include 30 tree species of which 17 are broad leaf, nine species of ferns, three of club moss and over 50 species of wild flowers. The flora is as it was in the early 1600s, a microcosm of the historical Appalachian forest. The trees include virgin hemlock up to 90 feet high and 21 feet circumference form cloisters in the park. A soft wood tree of this size has a carbon content of 1265 Kg[45] and a CO_2 equivalent of 4638

---

[43] Science Daily (July 9, 2012), "Climate in Northern Europe Reconstructed for the Past 2,000 Years: Cooling Trend Calculated Precisely for the First Time". ScienceDaily.com/releases/2012/07/120709092606.htm?

[44] Current models (7) used to forecast a hurricane's path in real-time is only sophisticated enough to generate a "fan" of probabilities for future path projections.

Kg. A virgin hardwood with an average diameter of 10 feet would have about 5542 kg of carbon and 20,322 kg of CO_2 equivalent locked up.

Research has found that for most species of trees the biggest trees increase their growth rates and sequester more carbon as they age. The international research group reports that 97% of 403 tropical and temperate species grow more quickly the older they get. The researchers[46] reviewed records from studies on six continents. Their conclusions are based on repeated measurements of 673,046 individual trees, some going back more than 80 years.

Earth's thermal history suggest that an increase in CO_2 within the influence of the Hydrologic Cycle (aka Water Cycle[47]) will help the fauna and flora thrive and result in no significant influence upon Earth's overall thermal budget when compared to the importance of water vapor and dust, the Sun's output and et cetera.

Summary- A new NASA study[48] shows that tropical forest may be absorbing far more CO_2 than many scientists thought, in response to rising atmosphere levels. The tropical forest absorb about 0.56% of the total global absorption while the northern forest takes in 0.44%. There is more CO_2 available at the equator because the CO_2

---

[45] One pound = 0.453592 kilograms.
[46] *U.S Geological Survey Western Ecological Research Center plus three Oregon State University researchers.*
[47] *The hydrologic cycle (water cycle) describes the existence and movement of water on, in, and above the Earth. Water on Earth moves continually through the hydrologic cycle of evaporation and transpiration (evapotranspiration), condensation, precipitation, and runoff. Its final destination is usually the oceans. Evaporation and transpiration contribute to the precipitation over land. The water cycle has been working for billions of years and life on Earth depends upon it continuing to work. (US Geologic Survey)*
[48] *NASA/Jet Propulsion Laboratory, "NASA finds good news on forest and carbon dioxide:, ScienceDaily, 2 July 2015*

molecule is heavier than air and centrifugal force from the Earth's rotation causes the molecule to migrate there.

Earth's climate is made up of an infinite family of over lapping micro/local/regional climates. How does the global climate system as a whole interacts, producing an influence upon the end result weather for any given period and location? What is a logical representative period that can depict a true climatic trend? Is it a century; millennium; or a geologic epic? The simple answer is that with the small historical data string and current state-of-the-art similitude techniques only allow for a cursor definition of what may happen now. The local/regional weather systems that accumulate into the global climate controls are the total that Earth's has contributed to the atmosphere column to date. The oceans chemical content is a testimony to that runoff resulting from the precipitation generated by the Water Cycle.

Satellite[49] data proves that the world was greener in 2010 than it was in 1982. There appears to be an 11% increase in foliage across parts of the arid regions of Earth because of CO_2 fertilization. This is another indicator that Earth's regional climates check and balance system works within acceptable parameters of about +/- 3°C. E.G.-A by-product of the Sahara Desert blooming, which it has before, is the Atlantic hurricanes will decrease significantly because of a lack of dust!

While flying for the military (1960-1972), I have, on occasion, observed a clear path with unlimited visibility created from the precipitation from a thunder storm as it passed through a hazy air layer in the lower atmosphere that had a visibility of less than one mile. Precipitation appears to be the cleansing agent for the lower air column. Thus, any gas produced on Earth that has a density higher than the atmosphere will not go beyond the influence of the **hydrologic cycle**[50] which produces all of Earth's precipitation.

---

[49] www.sciencedaily.com/releases/2013/07/130708103521.htm?
[50] Water exists on earth as a solid (ice), liquid or gas (water vapor). Oceans, rivers, clouds, and rain, all of which contain water, are in a frequent state of change (surface water evaporates, cloud water precipitates, rainfall infiltrates the ground, etc.). However, the total amount

Thus the CO_2 does not ascend but descends!

---

*of the earth's water does not change. The circulation and conservation of earth's water is called the **"hydrologic cycle"**.*

THE MECHANICS OF EARTH'S CLIMATE ENVIRONS

"The more you understand it, the more you realize you don't have to answer the question of whether or not something is conscious in order to define consciousness". But, he said, it's important not to come up with a definition before we've understood all the elements that need to be encompassed in that definition, least we suffer what he calls "the heartbreak of premature definition," an intellectual dysfunction he believes many of today's consciousness scholars suffer from.[51]

"WEATHER, n. The climate of an hour. A permanent topic of conversation among persons whom it does not interest, but who have inherited the tendency to chatter about it from naked arboreal ancestors whom it keenly concerned. The setting up of official weather bureaus and their maintenance in mendacity prove that even governments are accessible to suasion by the rude forefathers of the jungle" — Ambrose Bierce

Mark Twain famously observed: "The difference between the almost right word & the right word is really a large matter–it's the difference between the lightning bug and the lightning."

Introduction

"Subject: Global warming prediction
The Washington Post
The Arctic Ocean is warming up, icebergs are growing scarcer and in some places the seals are finding the water too hot, according to a report to the Commerce Department yesterday from Consulafft, at Bergen, Norway.

Reports from fishermen, seal hunters, and explorers all point to a radical change in climate conditions and hitherto unheard-of temperatures in the Arctic zone. Exploration expeditions report that scarcely any ice has been met as far north as 81 degrees 29 minutes.

---

[51] *Moffett, Shannon, "The Three-Pound Enigma", Algonquin, 2006. Quote is from Doctor Daniel Dennett, Philosopher*

Soundings to a depth of 3,100 meters showed the gulf stream still very warm. Great masses of ice have been replaced by moraines of earth and stones, the report continued, while at many points well known glaciers have entirely disappeared.

Very few seals and no white fish are found in the eastern Arctic, while vast shoals of herring and smelts which have never before ventured so far north, are being encountered in the old seal fishing grounds. Within a few years it is predicted that due to the ice melt the sea will rise and make most coastal cities uninhabitable".
* * * * * * * * *

I neglected to mention that the above report was from **November 2, 1922** as reported by the Associated Press and published in The Washington Post: 90+ years ago.

Date: December 1, 2014- Source: Michigan State University

The most comprehensive evidence to date that climate extremes such as record temperatures and droughts are failing to change people's attitudes about global warming are presented by research scientists. **Political orientation** is the most driving factor in shaping perceptions about climate change, both short and long-term, according to the research[52]. The influence that shifting climate patterns are influencing perceptions in the United States was not found to be in the study results, but they showed that politics has the predominate effect on climate change perceptions. More than 100 computer models were used to analyze a decade plus of Gallup survey responses on climate change with 50 years of regional climate data from the National Oceanic and Atmosphere Administration (NOAA).

The study also analyzed climatic storm-severity measures by the NOAA, e.g.-temperature, drought, precipitation and wind velocity

---

[52] *Sandra T. Marquart-Pyatt, Aaeon M. McCright, Thomas Dietz, Riley E. Dunlap,* **Politics eclipses climate extremes for climate change perceptions.** *Global Environmental Change, 2014;29:246 DOI: 10.1016/j.gloenvcha.2014.10.004*

from all 50 states and compared the data with 11 years of public opinion data.

With this in depth study of these issues there is "little grounds for optimism," the study says, "that public concern about climate change will be driven by future climatic conditions."

There is another study[53] that found that some scientists are tweaking experiments and analysis methods to increase their chances of getting results that are easily published". P-hacking is when researchers either knowingly or unknowingly analyze their data many times in many ways until they get the desired results. This pressure is driven by the fact that scientists are judged by the number of their publications and the importance of the scientific journals they go into. Dr Head said that some scientists adjust their experimental design, datasets or statistical methods until they get a result that crosses the significance threshold. They might look at their results before the experiment is finish or explore data with multiple statistical methods without realizing this can lead to a bias result.

According to a study[54] [55] funded by the National Science Foundation the researchers discovered that the abrupt climates changes affected Greenland first, with a response to the Antarctica climate delayed by approximately 200 years. From ice core documentation there were 18 abrupt climatic events that were documented during the last 68,000 years. The abrupt events of the ice age were regional in scope and probably tied to changes in ocean circulation. It appears that the oceans (heat sinks) are acting as thermal reservoirs and this mechanism is balancing the two hemispheric thermal budgets within the 200 year period.

[53] *Head, Megan; Holman, Luke; Lanfear, Rob; Kahn, Andrew; Jennions, Michael; "The Extent and Consequences of P-Hacking in Science", Australian National University, 2015*
[54] *Oregan State University, "200-year lag between climates events in Greenland, Antarctica: Ocean involved", ScienceDaily, 29 April 2015*
[55] *"Precise interpolar phasing of abrupt climate change during the last ice age" Nature, 2015*

Current climate models[56] are not able to include complete models of Greenland and Antarctic ice sheets and to document, reliably, how ice field changes affect sea levels. The models failed to accurately account for ice mass increases and losses from snowing and melting, respectively.

Some miss guided Cases

1. By 2014 California was suffering a severe drought. It appears that the hydrologic yield determination for the reclamation plan in California and the far west were based on a "wet" water cycle (early 1900s) and that "normal" appears to be extremely drier than predicted. It's like drilling water well on the side of a mountain which penetrates a "perched aquifer"[57] which has limited water storage. Over time the well goes dry because the aquifer inflow is less than its outflow.

This is a case where sophisticated misinterpretation of water yield and statistical regression procedures[58] allows the "advancement" of the technology e.g. modeling, etc.

2. Another example of applying highly technical procedures and modeling techniques and methodology to simulate at best, the random occurrences of natural events that are cyclic (frequency can be anytime; e.g.-from days to a millennium, etc.). Simplistic modeling techniques that are comparatively comprehensive, since it's the first time tools (calculators, slide rules, computers, etc.) were available which would allow gigantic number crunching. Yet the interaction of any process is so huge that depiction with today's "tools" is virtually impossible. Yet today's applications are more incomprehensibly advanced from one generation to another.

---

[56] Department of Energy, Office of Science, "New methods relates Greenland ice sheet changes to sea-level rise", ScienceDaily 14 April 2015
[57] A type of unconfined aquifer that sits above another unconfined aquifer because water infiltrating from the surface is trapped or 'perched' on a shallow aquitard. An **aquitard** is a zone within the earth that restricts the flow of groundwater from one aquifer to another
[58] Frequency analysis of the historical data to determine the probability of water availability over time, in other words what are the risks or chances.

Technology advancements are a spiral for almost every conceivable area of science, yet the immaturity of their application and the incomprehensible or misunderstanding of the results by the masses, yet the zeal of the scientists to be the "expert" or an explainer of the phenomena and receive the notoriety or award for being the innovator is over whelming. The definition of the "learning curve" is always in question. Yet the knowledge curve and the "real" understanding curve are "light" years apart but closing. As the two curves become relative close technology has advanced to the point where this particular knowledge is not germane to the overall scheme of how things interact.

All the above can be applied to the studies for: nuclear winter; oil spills; water vapor, dust; ozone; public polls; carbon dioxide; etc.

Comment: Some people base decisions upon the "last" collected data before a decision is rendered therefore polls are "predictable" because they are biased! Scientific results are often political results arrived by only using specific data streams to "prove" the foregone conclusions.

This begs the question: Is the public education of people depicted by the "learning curve real or imaginary?

Global climate(s) science continues to be held hostage by a debate between the scientific and special interest communities. Attempts by special interests groups have driven scientists to *simplify* an extremely complicated global weather system. To date, no entity has produced a simulation model that has been calibrated with any definitive precision to the degree of simulating an accurate projection, other than statistically based forecasts, for any single local point on Earth. Current forecasting has produced an attitude amongst non-scientists leadership causing the climatic debate to move away from credible scientific study to a "culture" that has changed the non-scientific answer to: "do you believe or not believe that "global warming" is happening?" Without any qualitative assessments!

Universities appear to teach science that fits the world that we have created, which is not the Natural world. In fact the "environmental movement" has shaped academia's curriculum in the natural sciences. Very few institutions, if any, teach science from the view point of "why" or "how". Few teach the analysis of ALL data that's possible to observe to arrive at a conclusion. The very opposite appears to be true. We know the answer so let's design our observations to collect data to verify our fore gone conclusions. Emphasis is placed upon the effect and not the cause. The following are some examples.

DUST: There is always a thin layer of dust circling the Earth in the upper reaches of the atmosphere. Beautifully colored sunsets are clues, especially after a volcanic eruption. Thousands of tons of dust are blown off Earth's deserts each year. Dust is probably the second most important regulators, after water vapor, of Earth's temperatures. Water vapors, which cannot occur in the atmosphere without dust particles, are the main players in climate control. In fact every drop of precipitation, rain, snow or ice, must have a speck of dust around which to form. Water vapor that forms clouds must have particles to form. Without dust the humidity below 300% will not condense. Above 300% humidity water will condense on objects including humans.

Dust in the atmosphere is produced by saltation and sandblasting of sand-sized grains and it is transported through the troposphere. This airborne dust is considered an aerosol and once in the atmosphere, it can produce strong local radiative forcing. Saharan dust in particular can be transported and deposited as far as the Caribbean and Amazonia, and may affect air temperatures, cause ocean cooling, and alter rainfall amounts.

Let's look at the record storms of 2011 that occurred in the south and Midwest of the USA. The thousands of wild fires burning in the southwestern part of the USA produced many tons of dust into the atmosphere. This dust contributed to the volatility of these storms.

No weather model was used in forecasting nor did the meteorologists presenting these forecasts mention, in their interpretations of conditions the possible quantitative impact, that dust, as an input, from these wild fires had any effect upon what was happening weather wise.

Without dust from the Sahara Desert, the Atlantic hurricane season would not exist. The Caribbean Islands would consist of grey rock without dust from the Sahara, which produced the Islands' layers of top soil.

Dust may turn out to be the most important thermal climate regulator of all the culprits that science has assigned that roll.

Everything on Earth, alive or not, is in a constant state of entropy, thus turning into dust. The millions of tons released into the atmosphere each year have an enormous effect upon Earth climates, from precipitation to becoming filament to absorb and retain calories/heat.

Misperception: Plastic bottles last longer in certified sanitary landfills than paper milk cartons: Everything in a properly designed land fill that was not once alive deteriorates at about the same rate over time. The chemical (carbon) makeup of plastic may break down before paper. Landfills have not been around long enough to determine their preservation lineage. Methane from decomposing once-alive compounds is the only thing that escapes a properly designed landfill.

Human activities affect global climate?

Paying attention to the climatic details by defining/creating a model simulation that can depict the interrelationships of all of Earth's regional climates has been lost in the debate. This lack of scientific "state-of-the-art" exercise to calibrate any model with ground truth data has caused non-technical folk to dominate the discussions. The continued gathering and analysis of meteorological knowledge seem to be futile exercises. Neither climatic knowledge nor its precise manipulation that is "cause and effect" driven gets added to a one-sided debate.

In an effort to present a definitive global system that is "cause and effect" driven, utilizing the known climatic components will begin to make some sense in the continuing debate. The reasons as to why global ice advancement is more likely to occur than any significant long term global warm up are also presented. Included is the scientific rationale as to why **Antarctica is Earth's thermostat**.

Scientists have discovered how tree roots in the mountains may play an important role in controlling long-term global temperatures. Researchers from Oxford and Sheffield Universities have found that temperatures affect the thickness of the leaf litter and organic soil layers, as well as the rate at which the tree roots grow. In a warmer world, this means that tree roots are more likely to grow into the mineral layer of the soil, breaking down rock into component parts which will eventually combine with CO_2. This process called "weathering" draws CO_2 out of the atmosphere and cools the planet. The researchers concluded that this theory suggest that mountainous ecosystems have contributed to Earth's thermostat, addressing the risk of 'catastrophic' overheating or cooling over millions of years.

Enhanced growth of Earth's leafy greens during the 20th century has significantly slowed the planet's transition to being hot, according to the first study to specify the extent to which plants have prevented climate change since preindustrial times. Researchers at Princeton University found that land ecosystems have kept the planet cooler by absorbing billions of tons of carbon, especially during the last 60 years. Earth's carbon-storage capacity has kept 186 billion to 192 billion tons of carbon out of the atmosphere since the mid-20th century. From the 1860s to the 1950s, land use by humans was a substantial source of carbon entering the atmosphere because of the deforestation and logging. After 1950s, however, humans began to use land differently, such as by reforestation and adopting agriculture practices that is higher yield yet on a larger scale. At the same time, industries and automobiles continued to steadily emit CO_2 that contributed to a botanical boom.

These "carbon savings" amount to a current average global temperature that is cooler by one-third of a degree Celsius. Earth has warmed by only 0.74 degrees Celsius (1.3 degrees Fahrenheit) since the early 1900s; compared to a mere two degrees Celsius which some scientists calculate would be the point that would be dangerously high. (Note: Earth's normal range of temperatures is plus or minus three degrees Celsius.) Changes in CO_2 emissions from land-use activities need to be carefully considered. Until recently, most studies would just take fossil fuel emissions and land use emission from **simple** models, which does not consider how managed lands such as recovering forest take up carbon. It's not just climate its people that are major drivers on land in the carbon game; they are not just taking carbon out of the land they are changing the land's capacity to store carbon! Models should be refined to incorporate these changes to improve their confidence levels.

Researchers at Kansas State University have found an upside to the relative higher CO_2 levels. Their research has shown that there is a mitigating effect that droughts has on winter wheat and sorghum by allowing more efficient use of water. Their data goes back to 1958 when the first accurate measurement of atmospheric CO_2 was made.

Earth's climate is an aggregate of regional climates working as a global system driven by orographics[59], plate tectonics, thermodynamics, astrophysics and kinetics. This is to say mountains, moving continents, deserts, plains, hot or cold liquid (air and water) interfaces, Sun, and moon, all provide a force that results in the coordinated movement en masse.

---

[59] *Highlands and mountain ranges*

The oceans are the "heat sinks" that keep the global temperatures moderate. Ocean currents, called conveyors or streams, carry the equatorial heat to the Polar Regions. The conveyors are guided by the continents and the Earth's rotation. If a conveyor is stopped, the applicable tropics will heat up and the cold latitudes will become colder. If a conveyor speeds up the heat distribution will lessen, thus temperatures, over time, will also fall in the Polar Regions.

Air is less dense therefore has minimal effect upon Earth's temperatures as compared to an equal volume of water. For example, you can spend more than 20 minutes inside a dry sauna at temperatures of 65°C (190° F), but you can't hold your hand under a faucet of hot water at 36°C (110°F) for more than a few seconds.

The globe is covered with distinct weather systems that overlap, interact and thus cover the entire globe. The Gulf Stream is one example of a weather system that affects a regional climate.

The regional climates consist of a series of distinct local climates. An example of a local climate is the Los Angeles Basin. This "bowl", created by the mountains, under certain weather conditions, can cause smog to form because of exhaust from a high concentration of automobiles. One weather system moving through an area can refresh/scrub the local climate.

Before several continuous scenarios of assumed rising temperatures can be played out, the major components such as ice, water, solar, celestial, internal radiation, state-of-the-art modeling, ocean conveyors, continent locations, etc., should be collectively described as they relate to Earth's climates. At best with today's technology, we can vaguely comprehend the sophistication of how an infinite number of micro climates could result, via the "butterfly effect", into multiple environments over time.

Before we can comprehend what ingredients constitute a global "climate", we must have an awareness of the magnitude, diversity and difficulty that this understanding may pose. We know that knowledge of a local climate cannot readily be extrapolated to a global one without the analysis of a millennium of data. To enter this data in today's modeling would be nearly impossible because of the modeling techniques. If it were done, the results would simply be unreliable! One example is the aggregate composition of models used to create a 'statistical fan' to describe the future track of a hurricane's path. Thousands of 'real time' data observations don't change the forecast from being parts scientific and experience.

The National Research Council arm of the National Academy of Sciences and National Academy of Engineering reported[60] that the United States' collection of climate models should be advanced substantially to deliver more detailed, smaller scale climate projections. The report suggest that climate models should take a more integrated path using a common software infrastructure while adding regional detail, new simulation capabilities, and new approaches for collaborating with their user community.

Short sightedness that ignores the magnitude of a global translation that requires extrapolation of local visual evidence (i.e.-retreating glaciers) as a global condition is like picking a single cell from your body to evaluate your body's health! We bias the results by "knowing" the answer before we collect and analyze the data to arrive at a conclusion.

Earth and the moon rotate around each other, as if tied together with a string, as they travel around the Sun in an orbit that has never been fixed. Earth is a gyroscopic sphere flying through space at more than 69,361[61] miles per hour. It tilts, wobbles, precesses and

---

[60] ScienceDaily.com/release/2012/09/120907125152.htm?utm
[61] Spin and orbit speed = 69,361 mph; Earth is moving toward Lambda Herculis at 43,200 mph; Earth's motion perpendicular to Galactic Plane = 15,624 mph; The Galactic spin rate =446,400 mph; and if you left the Milky

is affected by everybody in the Solar System and beyond. Since Earth was formed some 4.66 billion years ago, it has never experienced any moment like any other! For example, after the winter solstice "the position of the perihelion (nearest the Sun) shifts steadily and makes a complete circuit of orbit in 21,310 years. The actual amount of the tilt changes very slightly, growing a tiny bit more, then a tiny bit less, and in slow oscillation. All of these changes have a small effect upon Earth's average temperature, not great, but enough at certain times to pull the trigger for either the advance of glaciers or their retreat.[62]"

900 million years ago Earth's day was 18 hours[63] long and a year[64] w as 481 days duration. The effect of the moon has caused the changes that are today's conditions. Future days will become longer and the years will become shorter. Future building of wind generators will exacerbate the slowing of Earth's rotation since this is the energy used by the generators to produce electricity.

Earth's surface is a water domain. Water covers about 71% of the surface yet it constitutes only about 1/4200th of Earth's total mass. Oceans hold about 97.2% of Earth's water and are the source for fresh water to the tune of 80,000 cubic miles evaporated each year that fall in the form of rain or snow. Stored underground are some 200,000 cubic miles of water, mostly fresh, with an additional 30,000 cubic miles stored in lakes and rivers.

Water, in a solid state, covers about 10% of the Earth's surface which is roughly the size of the North American continent. The Antarctic ice sheet contains about 91% of the total ice on Earth. Greenland has about 8% of Earth's ice while the mountain glaciers and Arctic cap account for less than 1% of the total.

Way Galaxy add another 1,339,200 mph, the speed the Milky Way is moving through the Universe.
[62] Asimov, Isaac, "New Guide to Science", Basic Books, 1984
[63] The time it took for one complete rotation.
[64] The time it took for one orbit around the Sun.

The following scenarios are based upon an assumption of continued rise in Earth's temperatures to the point where cooling will obviously begin in most regions.

Arctic's ice cap floats. Its position at any given moment is at the whim of the ocean conveyors, continental boundaries and prevailing winds. The melting of the Arctic ice will be from calories given up by the ocean and not the atmosphere. Since this ice total is less than one per cent of Earth's ice the effects, if melted, are local only. The ebb and flow of these ice packs are a result of a regional climate, not global.

If **ALL** the ice from the Arctic and mountain glaciers melted at the end of the summer (northern hemisphere) the effect upon sea level would be negligible, nearly impossible to ascertain since the oceans of the world are subservient to wind fetches and tidal gyrations from a few feet to many tens of feet. However, there is evidence that this fresh water melt, augmented by the Greenland melt, did overlay the cold Arctic salt waters to the point that it shut down the Gulf Stream (GS) for 10 days in 2004. This indicates that this regional climate is very fragile from a thermal observation.

No scientist knows what caused the GS to stop flowing in 2004. According to the Scientists at Woods Hole, the stoppage event was described as "the most abrupt change in the whole (climate) record".

During the recent severe winters of 2009 and 2010 in Europe the National Oceanography Center measured a strong reduction in the strength (ability to convey calories northward) of the GS. This caused these European winters to emulate the winters in western Canada at similar latitudes.

What would happen if "a significant amount of Greenland's ice cap melted"?

The latest climate models predict that the GS will slow down as global warming increases. However, measurements by NASA[65] of the Atlantic Meridional Overturning Circulation (GS) show no significant slowing over the last 15 years; in fact the data suggest the circulation may have sped up by as much as 20% in the recent past.

The GS is the conveyor that keeps England and northern Europe from having a regional climate similar to the climate in Canada above the 45th latitude.   Warm surface water flows from the tropics northward into the North Atlantic as one of the currents that make up the Atlantic overturning circulation system.   Within the oceans surrounding Greenland the GS cools and sinks to great depths as it changes direction.   What was once warm surface water heading north becomes cold deep water heading generally south ward.

The GS starts from the Equatorial Current from the African coast, moving east to west under the influence of the trade winds in the tropical North Atlantic.   The South American continent deflects the current northward causing it to meander among the Caribbean Islands.   The Equatorial Current circles the Gulf of Mexico in a clockwise fashion, exiting through the straits between Florida and Cuba.   Then the Stream joins the Antilles Current, officially forming the GS.

The GS is about 90 kilometers wide and flows at two meters per second at about 60 degrees latitude.   The GS flows at about 80 million cubic meters per second, which exceeds the volume of ALL rivers in the world.   The volume of the GS is 3500 times larger than the Mississippi River's discharge into the Gulf of Mexico.

The large volume of warm water moved by the GS toward the colder North East Atlantic reaches near Latitude 40-42 degrees north before it's deflected southward.   The GS loses heat energy by melting the ice floes, as well as calories loss as the cold fresh water, from the glaciers, all of which overlay the GS cooling it to a density[66]

---

[65] Jet Propulsion Laboratory, Pasadena, CA Press release dated March 25, 2010
[66] Density of the cold salt water = 1.030+; Ice melt density = 1.000

of the surrounding salt water. The result is the GS loses its identity and becomes part of the North Atlantic Ocean.

If melt from the Greenland ice pack increases, there will be an increase of fresh cold water with a density of 1.000 overlaying the ocean of cold saltwater with a density of 1.030. The boundary[67] integrity of any two liquids of different densities is very rigid.

Greenland's ice cap volume is about 2.85 million cubic kilometers. If all the ice melted, the mean elevation of the world's oceans would be increased by about 23.6 meters. But, more realistically, for each 100 meters of ice melt equivalent to the Island size, the oceans would rise about 19.5 inches.

Since Greenland on average is warmer than Antarctica, an increase in local temperatures could produce melting here first. If the temperatures on Greenland continue to rise, the snowfall will increase on the ice cap. This will increase the ice cap volume and provide more ice for glacier calving into the North Atlantic. This is why it's uncertain if the ice sheets on Greenland and Antarctica are growing or shrinking. Antarctica is so cold that surface melting will not occur, but Greenland is a different story by 50+ degrees Fahrenheit.

Once the GS is cut off, the regional climates of northern Europe will no longer be the recipient of the tropical heat energy. With time, their climate will emulate that of Canada above the 45° latitude. The Polar ice cap will grow to include the North Sea and will attach itself to the continent. Ocean currents, which are the conveyors of surface energy around the globe will be modified, energy wise, to the point that the Arctic ice cap will, over time, expand. As the ice cap grows heat from the Sun will be deflected and the ice pack will continue to enlarge. Greenland's seasonal temperatures will start to

---

[67] *The boundary is called a thermocline.*

decrease, the ice pack will begin to grow and the warm-cold cycle will continue as it has historically.

Northern Europe and the North Sea oil platforms will become uninhabitable over time. The increase in ice coverage will reflect the Sun's energies during the summer, and the Earth will begin a cool cycle. The increase in ice coverage will take water from the oceans thus they will recede.

A group of scientists from the Institute of Geography at Johannes Gutenberg University Mainz[68] (JGU) used tree-ring density measurements from sub-fossil pine trees from the Finnish Lapland to reconstruct the regional climate back to 138 B.C. The researcher's precise results for the temperature data for the last two millennia has been a climatic cooling of a minus 0.3° C (centigrade) per millennium. They conclude that the cause was a gradual change in the position of the Sun and an increase in the distance between the Earth and the Sun. The researchers produced a detailed graph of the temperature variations which show a +3° to a -3° C fluctuation every decade or so. These changes in the amount of thermal energy were delivered by the GS.

? What regional climate changes will result from a continued warming trend, say a normal four to six degrees Centigrade for the next two centuries

- Here's my list:
- *Antarctica's annual precipitation would nearly double over time taking moisture from the oceans.*
- *New York would become a river city.*
- *The corn and wheat belts of Iowa and Nebraska would slowly move northward to the plains of Alberta and Saskatchewan.*

---

[68] *www.nature.com/nclimate/journal/vaop/ncurrent/full/nclimate 1589.html (published on line 08Jul2012)*

- *All continents would gain land mass from the oceans receding exposing continental shelf as the ocean would fall nearly 450 feet in a couple centuries. The Florida Keys would become a peninsula.*
- *Greenland's ice pack would begin to grow as the glacier movement slows and calving of ice also slows.*
- *Ice sheet covering the South Pole would expand at an alarming rate.*
- *Southern oceans cluttered with ice floes would cause water temperatures to drop significantly.*
- *Antarctic's research station would be abandoned because of ice movement caused by an increase in precipitation.*
- *Annual precipitation in the USA's upper Midwest would double to 20+/- inches. This increase in runoff would, over time, cause the Mississippi River's annual flow to gradually increase to the point that diversion around New Orleans would become necessary only during major flood events. As the river channel eroded because of the receding Gulf levels, New Orleans would eventually be above sea level.*
- *Oceans would cool three to five degrees C. This would cause the air temperature to cool. Local areas would experience ice ages which would shorten the summer season. The polar bears would return to the ice.*
- *If the precipitation in Antarctica doubles, the moisture must come from the oceans. Since the continent is so large an additional inch over this area for two hundred years constitutes a large volume of water.*
- *The Southwest US is nearly a desert now so a very little rise in temperature would cause the local climate to continue in that direction. Since the weather pattern in the US travels from the Southwest toward the northeast, there would be more dust available to create an increase in precipitation down range. Thus, weather systems in the US may become more volatile.*
- *Population in Egypt would expand beyond the Nile Valley as the desert blooms.*
- *The Atlantic hurricane season would nearly cease as the Sahara desert blooms.*

- *If the Sahara Desert blooms, there would be less dust available to form clouds and/ or rain drops. If a hurricane does not have clouds and/or rain to dissipate its energy, will it self destruct?*

Increasing regional temperatures over time would cause the global temperature to cycle to a cool down. There are several "safeguard" mechanisms in place to prevent a warm up, but none to prevent an ice age. During Earth's history, Antarctica has proven to be its thermostat. Earth has never been overheated since the initial cool down and as long as the thermostat is in place, it will not be.

Before the study of plate tectonics became a science, some people as early as 1596[69] believed that the arrangement of continents appeared to be puzzle pieces that could have fit together to form a super continent in years past. Currently the "theory of continental drift" suggests that some 225 to 260 million years ago all seven continents were together forming a super continent called "Pangaea"[70] . For more than 225 million years, the Antarctic continent has remained near Earth's "bottom" while the other continents, North and South America, Europe, Asia, Africa and Oceania (Australia), have drifted[71] to their present positions. Evidence from oceanic ridges surrounding Antarctica indicates that the super continent began to break up about 150 million years ago. Fossils, soil, rock, modeling and other evidence support these conclusions.

---

[69] *In Dutch map maker Abraham Ortelius' work "Thesaurus Geographicus" he suggested that the Americas were "torn away from Europe and Africa by earthquakes and floods". It was not until 1912 when the idea of moving continents was seriously considered as a scientific theory called "Continental Drift", introduced by a German meteorologist Alfred Lothar Wegener.*
[70] *USGS (U.*
[71] *Today these continents continue to drift at the about the same speed that our finger nails grow. (US Geological Survey USGS) The "theory of plate tectonics" states that **Earth's** outer most layer consist of a dozen or more plates that are moving relative to one another as they float on hotter, more fluid material.*

The Antarctic, on average, is the coldest, driest, and windiest continent, and has the highest average <u>elevation</u> of all the continents. Antarctica[72] is the fifth largest continent by land mass, and the third largest, if the area of the ice cap is measured, assuming that Greenland is part of the North American continent which is the second largest. The summer temperatures rise to about minus 30°C to Earth's lowest yearly air temperature of about minus 80°C. The relative humidity is 0.04 % (Mars' is 0.03%) making the continent the driest place on the globe. Annual precipitation averages just above an inch, mostly from ice fog over the interior regions.

---

[72] *Winkel-tripel-projection.jpg*

PERMIAN
225 million years ago

TRIASSIC
200 million years ago

JURASSIC
135 million years ago

CRETACEOUS
65 million years ago

PRESENT DAY

Plate tectonic history (from Bing.com)

Earth's rotation creates a gyroscopic effect that has appeared to lock Antarctica in its current position at the "bottom" of the globe. Forces from a gyroscope react along a three dimensional axis, separated by 90 degrees; e.g. spin, output and input. Earth's spin axis is the global center. The Sun provides the input, and the output axis is 90° from the input. Thus it affects the South and North Poles. The continent lies 90° from the centroid of a gyroscope, Earth, and its gravity "string" from the Sun. The effect of the forces created by the continental drift appears to be inhibited by this gyroscopic effect. *Or is it something simple like the Antarctic continent continues to float on the apex of a globe that is spinning!*

95

Furthermore, Antarctica's position at the apex of a spinning sphere appears to have a safeguard against thermal increases. According to research[73] results the "*Antarctic Circumpolar blocks the southern Hemisphere equivalent of the Gulf Stream from delivering heat to Antarctica.*" The Antarctica continent has been in this position, tectonically, for more than 260 million years, thus it's not going to move in the near term, geologically speaking.

During the Triassic and Jurassic periods, 200 to 150 million years ago respectively, when most of the ingredients for fossil fuels were laid down, the dinosaurs grew to be very robust because of an abundance of O_2 in the atmosphere. The Earth warmed up a bit, as its thermostat worked perfectly. Had Homo sapiens been around at that time, they too would have been more robust because of the abundance of oxygen.

Therefore, Antarctica has been in place for more than 225 million years and will remain as Earth's thermostat for many more millions of years, preventing future significant global warm-ups.

In May 2009 and February 2011 snow fall partly offset the recent ice loss from the Antarctica ice field[74]. The snow event in 2009 alone added approximately 200 gigatons (1.1023 billion short tons) of mass which offsets about 15% of the most recent 20 year ice sheet mass lost loss.

These snow storms were caused by a phenomena called "Atmosphere Rivers". These "rivers" stretch over several thousand

---

[73] *University of South Carolina, "Ocean currents: Debut of global mix-master: The Antarctic Circumpolar Current begins its eastern flow through the Southern Oceans 30 million years ago after the Tasmanian gateway, migrating northward tectonically, aligned with the mid-latitude westerly wind band." Sciencedaily, 25 August 2015.*
[74] *Gorodetskaya, Irina; et al; "The role of atmosphere rivers in anomalous snow accumulation in East Antarctica", Geophysical Research Letters, 2014*

miles of oceans transporting large amounts of moistures around the planet. These "rivers" are notorious for producing flooding in Europe and the Americans. Their importance to Earth's sea levels is only beginning to become understood.

The following highlights one theme from my previous list that deserves a further note:

In the early 1600's when Europeans first arrived upon these shores, the Appalachian Mountain range was covered with mature trees (virgin forest). The Appalachians run from Maine to Georgia. Most of the "virgin" timber has been harvested while some second growth forest has been preserved as National Forest. Hardwood forest in these forests will take from 150 to 500 years to develop old-growth characteristics in one or two generations of trees.

Cathedral State Park in West Virginia consists of only 133 acres of over 170 species of vascular flora that include 30 tree species of which 17 are broad leaf, nine species of ferns, three of club moss and over 50 species of wild flowers. The flora is as it was in the early 1600s, a microcosm of the historical Appalachian forest. The trees include virgin hemlock up to 90 feet high and 21 feet circumference form cloisters in the park. A soft wood tree of this size has a carbon content of 1265 Kg[75] and a CO_2 equivalent of 4638 Kg. A virgin hardwood with an average diameter of 10 feet would have about 5542 kg of carbon and 20,322 kg of CO_2 equivalent locked up. The large oak tree next to my door has about 15,000 kg of oxygen locked up that, if released, would make Homo sapiens more robust.

Did you ever wonder how the dinosaurs evolved to be so large? Or why so much of today's fossil fuels originated during this same period? The flora and fauna grew so lush because carbon dioxide was available in the atmosphere. During this time the growth rate

---

[75] *One pound = 0.453592 kilograms.*

excelled causing huge critters to evolve because oxygen was in the air!

Increasing local temperatures over time will cause the global temperature to cycle to a cool down. During Earth's history there have been safeguards against warming, Antarctica being the thermostat. Earth has never been overheated and as long as the thermostat is in place it will not be.

We as a scientific society have learned to construct devices that can perform measurements with extreme relative precision. Yet with today's technological knowledge and computational tools, we can only define weather in terms of probability or chaos theories. Our scientific advancements have been exceptional during the last 400 years, but we still have only begun to understand. We still have a few millennia before we can begin to create a model that can, with precision, emulate the "butterfly effect", therefore allowing us to make a precise forecast for a specific point. Once this milestone is reached, then we will have begun to advance. To date Homo sapiens's effect upon the globe can only be measured locally and NOT regionally or globally. Our arguments should be directed toward the sophistication of the science that defines the climate of our environment.

The virtual world that we have created, its causes and effect consequences, is more philosophy driven (special interest) than Nature driven. One group of "special interests" verses another group with a conflicting interpretation has created a near "virtual" reality of how Nature works. Unfortunately the debate is driven by non-scientists leaders who have determined the objectives and "cherry" pick the data to support the desired result. How does this all happen?

The virtual world that we have created makes no one accountable for the group's actions. Sometimes the solution is worse than the cure. For example take the wild fires that swamped the southwest during

2011. Years ago timbering was halted in the National Forests because an owl's habitat would be affected. No degree or significant was place upon a scientific analysis to quantified the degree of harm that selective timbering would accomplish. Emphasis was placed upon the group that would profit from the endeavor. This emphasis reflected the totality of the issue. But the lack of harvesting selected trees over time causes a buildup of fuel on the forest floor. Thus the result is a fire that is difficult to control, and the ultimate goal of protecting the owls is destroyed. There are many instances of similar disasters created by a "cure".

We are who we are by accident of birth and we quickly bond with our "tribe". This is human nature. The success of sports teams, armies and "causes" depends upon the bonding of the individuals. They "jelled" as a collective whole. These "team" members are considered to be the individuals that perform "inside" the envelope. Each represents what is expected in this world that we have created. Only the individuals that survive on the outer boundaries of the envelope, or even outside the envelope (beyond societal acceptances), make a contribution or advance the science of "being". These individuals probe the secrets of how Nature works. Thus they are not bound by artificial rules of the "world" that society has created.

This is my objective; to arrive at a probable effect based on the knowledge that we have accumulated to date with no preconceived notions or results.

A Summary of Global Climate History

"...the Earth was evidently coming out of a relatively cold period in the 1800s so that warming in the past century may be part of this natural recovery." **Dr. John R. Christy** *(leading climate and atmospheric science expert- U. of Alabama in Huntsville)*

Natural forces have always caused the climates on Earth to fluctuate. Periods of Earth warming and cooling occur in cycles. These fluctuations are called Milankovich[76] cycles. This is a well understood phenomenon as is the fact that small-scale cycles of about 40 years exist within larger-scale cycles of 400 years, which in turn exist inside still larger scale cycles of 20,000 years, and so on.

The Earth's axis completes one full cycle of precession approximately every 26,000 years. At the same time, the elliptical orbit rotates more slowly. The combined effect of the two precessions leads to a 21,000-year period between the astronomical seasons and the orbit. In addition, the angle between Earth's rotational axis and the normal to the plane of its orbit (obliquity) oscillates[77] between 21.5 to 24.5 degrees on a 41,000-year cycle. It is currently 23.44 degrees and decreasing.

Today a difference of only about 3 percent occurs between aphelion (farthest point from the Sun) and perihelion (closest point to the Sun). This three percent difference in distance means that Earth experiences a six percent increase in solar energy received in January than in July. This six percent range of variability is not always the case, however. When the Earth's orbit is most elliptical the amount of solar energy received at the perihelion would be in the range of 20 to 30 percent more than at aphelion. These changing amounts of received solar energy around the globe does result in prominent changes in the Earth's climate and glacial regimes. At present the orbital eccentricity (almost a circle as opposed to an ellipse) is nearly at the minimum of its cycle.

---

[76] Milankovitch theory describes the collective effects of changes in the Earth's movements upon its climate, named
after Serbian geophysicist and astronomer Milutin Milanković, who worked on it during his internment as a POW during the First World War. Milanković mathematically theorized that variations in eccentricity, axial tilt, and precession of the Earth's orbit determined climatic patterns on Earth through orbital forcing.
[77] Tilt that causes the seasons.

Earth's climates begin to rise about 11,700 years ago, rising from the **Pleistocene Ice Age** when much of North America, Europe, and Asia lay buried beneath great sheets of glacial ice. The **Pleistocene Ice Age** was a time when Earth's climate was dominated by ice ages and glaciers for the previous 2,588,000 years.

Pleistocene climate was marked by repeated glacial cycles in which continental glaciers pushed to the 40th parallel in some places. It is estimated that, at maximum glacial extent, 30% of the Earth's surface was covered by ice. In addition, a zone of permafrost stretched southward from the edge of the glacial sheet, a few hundred kilometers in North America, and several hundred in Eurasia. The mean annual temperature at the edge of the ice was −6 °C (21 °F); at the edge of the permafrost, 0 °C (32 °F).

Each glacial advance tied up huge volumes of water in continental ice sheets 1,500 to 3,000 meters (4,900–9,800 ft) thick, resulting in temporary sea-level drops of 100 meters (300 ft) or more over the entire surface of the Earth. During interglacial times, such as at present, drowned coastlines were common, mitigated by isostatic (reaching equilibrium) or other emergent motion of some regions.

The slight increases in climate temperatures during this current interglacial warm period, though a "normal" climate phenomena, have caused some alterations in the environment and the distribution and diversity of all fauna and flora.

Some examples:
1. About 11,700 years ago the Earth had warmed sufficiently to halt the advance of glaciers, and sea levels worldwide began to rise.
2. By about 8,000 years ago the land bridge across the Bering Strait was under water, cutting off the migration of men and animals to North America from Asia.
3. Since the end of the Ice Age, Earth's temperature has risen about 16 degrees Fahrenheit and sea levels have raised a total of 300 feet! Forests have returned where once there was only ice.

Over the past 750,000 years of Earth's history, Ice Ages have occurred at regular intervals, of approximately 100,000 years each.

"If 'ice age' is used to refer to long, generally cool, intervals during which glaciers advance and retreat, we are still in one today. Our modern climate represents a very short, warm period between glacial advances." **Illinois State Museum**

During the last 3 million years glaciers have at one time or another covered about 29% of Earth's land surface or about 17.14 million square miles (44.38 million sq. km.).   Beneath ice lay largely cold and desolate desert landscape. During the Ice Age summers were short and winters were brutal. Fauna and flora had a very tough time surviving.   Because of global warming Earth's temperatures have temporarily been moderated.

In the 1970s concerned environmentalists like Stephen Schneider of the National Center for Atmospheric Research in Boulder, Colorado feared a return to another ice age due to manmade atmospheric pollution blocking out the sun. Since about 1940 the global climate did in fact appear to be cooling. However, in late 1970s the temperature declines slowed to a halt and ground-based recording stations during the 1980s and 1990s began reading small but steady increases in near-surface temperatures. Fears of "global cooling" then changed suddenly to "global warming,"-- the cited cause: *manmade atmospheric pollution causing a runaway greenhouse effect.*

References

(1) *A scientific Discussion of Climate Change*, Sallie Baliunas, Ph.D., Harvard- Smithsonian Center for Astrophysics and Willie Soon, Ph.D., Harvard- Smithsonian Center for Astrophysics.

(2) The Effects of Proposals for Greenhouse Gas Emission Reduction; Testimony of Dr. Patrick J. Michaels, Professor of Environmental Sciences, University of Virginia, before the Subcommittee on Energy and Environment of the Committee on Science, United States House of Representatives

(3) Statement Concerning Global Warming-- Presented to the Senate Committee on Environmental and Public Works, June 10, 1997, by Dr. Richard S. Lindzen, Massachusetts Institute of Technology

(4) Excerpts from "Our Global Future: Climate Change", Remarks by Under Secretary for Global affairs, T. Wirth, 15 September 1997. Site maintained by The Globe - Climate Change Campaign

(5) Testimony of John R. Christy to the Committee on Environmental and Public Works, Department of Atmospheric Science and Earth System Science Laboratory, University of Alabama in Huntsville, July 10, 1997.

(6) *The Carbon Dioxide Thermometer and the Cause of Global Warming*; Nigel Calder, -- Presented at a seminar SPRU (Science and Technology Policy Research), University of Sussex, Brighton, England, October 6, 1998.

(7) *Variation in cosmic ray flux and global cloud coverage: a missing link in solar-climate relationships*; H. Svensmark and E. Friis-Christiansen, Journal of Atmospheric and Solar- Terrestrial Physics, vol. 59, pp. 1225 - 1232 (1997).

(8) *First International Conference on Global Warming and the Next Ice Age*; Dalhousie University, Halifax, Nova Scotia, sponsored by the Canadian Meteorological and Oceanographic Society and the American Meteorological Society, August 21-24, 2001.

(9) Ice Core Studies Prove CO_2 Is Not the Powerful Climate Driver Climate Alarmist.

TIME

"Time is Nature's way of keeping everything from happening at once." John Archbold Wheeler
"The reason that time has a direction is because the Universe is full of irreversible processes." Sean Carroll[78]

This fleeting instant, we term the present is the boundary that separates the future from the past. Time is not continuous but flows in increments that are so minute one cannot readily imagine. The dimension of this lull we call the present is the only moment when time stands still. Thus the "present" is of little significant to reality, it's simply a boundary between the past and future.

The two questions that have bugged scientists since the "dawn of science" are: (1) How did life start and (2) What is consciousness? Each leads to: What are we? How did we get to this place? Are we alone? Etc., etc., etc...

Why do I exist now? Why now rather than some other time? Why is it that the 20th century the "now" instead of some other time such as the 27th century or in Roman times? What's special about "now"?

If my mother had not met my father I would not have been borne nor would I have been born at some other time to other parents, but, when? Would this person be me? The flow of time presents so many unanswerable questions.

I have analyzed and researched all the data that I could mentally ingest in the last 60 years and I have arrived at a conclusion that many before me have made. "We don't even know what is possible to know!" The passage of time appears to be continuous in the world that we live in. Its measurement always depends upon one physical system relative to another. For instance, how many times does a clock hand rotate relative to the rotation of the Earth? These events are not related nor does one cause the other. Many time

---

[78] *"From Eternity to Here", Dutton, 2010*

measuring devices do not reflect the natural world. Since the bits of time are so small that we cannot measure them with today's technology, we have attached an arbitrary rate of one second per second to the flow of time. We record the rotation of the Earth as a day, divided into 24 hours; the time it takes the Earth to rotate around the Sun as a year yet neither measurement is precise thus the need for a leap year every four years to reset and make the system work. Furthermore day and year measurements are not relative nor do they depend upon one another in any way. Thus time measurements in this world that we have created are both relative and each arbitrary.

Researchers at the National Institute of Standards and Technology have achieved a level of precision in the strontium clock that's now accurate to a second over 15 billion years. These precision aspects have a broad impact on advanced communications and positioning technology, and upon the sensitive altimeters that are based upon changes in gravity and experiments that explore the quantum correlations between atoms.

If only we could anticipate the future. At family reunions I spend time catching up on the lives of kin folk that I have not seen or heard from in a year, and sometimes longer. Many times as we get older I am saying goodbye to kin without knowing it. In every case there are many family related thoughts that go unsaid. It is a reminder that for the living time is the cruelest dimension of all yet without time there would be no past or future. Thus without time there would be no memories, no good old days to reminiscent about, no myths to pass on from generation-to-generation.

Time is the integral part that defines our daily lives. We describe our activities around time. For instance what did we do last week or month or what are we planning for tomorrow all involve the past or future time. We identify and celebrate special events by "moments" in time but we despise its passing. Our hopes and fears are labeled in terms of time. "Do you remember the time that so and so did…?

"Time is our fourth dimension. Length, width and height make an object appear solid but time makes it real in appearance. For instance you are sitting in a specific desk in a particular room of a school building, which is described by an address and room number at a certain time of day, say two o'clock in the afternoon. At three o'clock you are still in the same seat but the past has eaten up another hour of the future. At some later[79] time you have moved on therefore you are no longer in the chair, so time defines your location, past, present or future.

Why does time seem to slow when you are waiting in line but fly by when you are having fun? You can lie on a beach for hours but become frustrated in just a few minutes in line. Why? Researchers have identified five elements that need to be in balance for people to experience the flow of time evenly: technology, people skill, plans and moods, rules and regulation, and cultural understanding. If these five elements are not in tune time will not seem to flow at the right speed and people will experience rush or drag.

We know that the arrow of time always moves forward. That is to say that the past always eats into the future at a rate of one second per second. What you see when you are talking to a person is that person in the past. The time it takes for the light image and the sound to travel from that person to you puts them in your past. Thus nothing you are aware of is in your present time. By definition the "present" is impossible. The adage that we are always living in the past is absolutely true; otherwise it would be impossible to communicate.

You are driving down a country road. When you look through the windshield you are looking at where you will be in a minute or so. Thus you are looking into the future. On the other hand if you look through your rearview mirror you see what you just passed a few

---

[79] *I never understood why the term "later" should define a future event.*

moments before. This is your immediate past. In this scenario you can see the future and the past so they both must be real.

All three time dimensions, past, present and future have always existed. The definition of time demands it. For example your birth date is in the past and since you exist the date must be real. Since you will die in the future your death date does exist; so do your future birthdays, anniversaries, etc. You are moving toward these dates at a rate defined by time. So your future does exist. These points in time have always existed. On the contrary there is a quote from Einstein: "The past, present and future are illusions". The definition of "illusion" could be "a figment of our imagination".

Every entity in space–time has a "base" from which time can only move forward to the future. This eliminates the possibility for any paradoxes. Some physicists believe that the past and future exist in a space-time continuum. Simply stated this means that the four dimensions, length, width, height and time define an event such as your birth and its position in a space-time continuum. This is to say that your birth's past still exists somewhere in the space-time continuum. If so, it would be located at a distance from you equivalent to your age times the speed of light.

Since we don't even know what is possible to know there is evidence that's beginning to emerge which implies the laws of nature, which we have discovered and defined by our technological theories, are changing with time. Since the concept of time is what distances eternity from the beginning these theories will no doubt change as we gain a better understanding of the natural world. I believe the flow of time is the catalyst that takes reality from the present to the future and provides the only avenue in which the past can be quantified. Since time is merely the rate of change from one spatial configuration to another, it is the fourth dimension.

Time allows one to journey across the knowledge world unabated by physical obstacles!

ATOMS ARE IDENTIFIED BY AN INTERNAL CLOCK

"By convention there is color, by convention sweetness, by convention bitterness, but in reality there are atoms and the void,' announced Democritus (460 – 370 BC. The universe consists only of atoms and the void; all else is opinion and illusion. If the soul exists, it also consists of atoms."

Gravity *is only the bark of wisdom's tree but it preserves it" Confucius (551-479 BC)*

The Periodic Table is a tabular display of the 118 known chemical elements organized by selected properties of their atomic structures. Thus the Table organizes the elements based upon how they perform not how they were formed. After years of research I have come to the conclusion that the "flow of time" inside the nucleus of an atom distinguishes each atom as it was transformed from the Hydrogen atom (H_2). This difference is in the state of matter, energy vs mass. Thus the catalyst that converts all matter into mass, from an energy state, is a minuscule increase in the applicable local time.

All atoms appear to contain the same amount of matter as the hydrogen atom, but each is identified by the different ratio of energy vs mass that make up this matter. The time clock inside the nucleus determines that ratio thus identifies each atom/element as different. Thus the H_2 atom is the progenitor of all atoms. That is to say that the H_2 atom plus time equals all the remaining atoms.

The secret to the definition of time could be revealed if we could discover how the atom was originally formed. The collecting, combining and "jump" starting particles to function as an atom is the question that needs defined. All atoms, born now in an expanding Universe, and in the beginning, are hydrogen atoms which appear to be the basis for ALL atoms.

Atoms are nature's storehouse for energy. They are the perfect battery. It is believed that 99.99% of an atom's mass is stored in its nucleus. Because of this stored energy the time zone within the nucleus is slowed to near zero because of the strong force. This puts the atom in a time warp with its outside environment. This could be one reason that a hydrogen atom, formed just after the Big Bang (BB), continues to exist in its original state after billion of years. Within this atom's nucleus a second is an eternity!

All atoms after hydrogen are simply modification of the hydrogen atom. Within the hydrogen atom the nucleus' matter is mostly in the form of energy because of a near zero time rate that exist in the nucleus as a result of being formed at the time of the BB. As the hydrogen atoms internal clock is altered by a cataclysmic event, i.e. a supernova or simply a star burning through its life cycle, more energy from its nucleus is converted into mass (particles). The "changed" atom takes on a different set of characteristics. Each time the new atom is subjected to this violence its mass grows and energy decreases as it becomes still a different atom. Supernovae are important in the nucleosynthesis of heavy elements. The explosions take a few tens of milliseconds creating selenium, copper, zinc and all the heavy elements that are found on Earth, and in our bodies. Synthetically, we have created elements that cannot survive in our time zone. Recently (2009) an element number 112 was created which is 277 times heavier than hydrogen. No. 112 resides on top of the Periodic Chart. This element was formed when zinc and lead were fused. This so called super heavy element existed for only a tiny fraction of a second before decaying, radioactively, into the elements that can exist within our time rate.

Hydrogen appears to be the "lightest" of all atoms because the matter of its nucleus exists almost entirely in the form of energy as oppose to mass. Thus the hydrogen atom has the greatest potential to be the most powerful of all atoms. Why gravity has the appearance of having a lesser effect upon energy compared to mass, within the atom's nucleus, is probably because of the strong force, a gravity component that bonds the nucleus.

Fusion[80] allows the harvesting of energy from an atom. This appears to occur when we manipulate the internal clock of certain atoms. An example is the energy released by causing a nuclear explosion by abruptly increasing the atoms internal clock, for an instant, to one second per second, the present time rate.

According to Fuller chemistry[81] is seen as "applied physics", whereas chemists claim their science is of things perceived by the human senses. From a physicists point of view it's conceivable that all atoms are identical in total "matter" but the internal structures of their nuclei are altered by virtue of the distribution of energy vs. mass. The difference in their respective uniqueness is the internal clocks-that-tick within the nucleus?

From a chemists point of view the **Periodic Table** is an orderly arrangement of known elements. It doesn't explain **how** the elements were created or what makes them similar but not the same. As we move up the **Periodic Table**[82], from hydrogen, the infinitesimal increases in the flow of time, within each atom's nucleus, cause the release of energy that allows certain particles, equivalent to that specific time, to be manifested, thus modifying the internal structure of the original hydrogen atom, creating a "new" atom. The new atom's atomic weight would be heavier because of the number of particles appearing as mass within the atom.

Energy appears to not be affected by gravity thus it presents no weight. We must conclude that elements that are created by

---

[80] Nuclear fusion is the process by which two or more atomic nuclei join together, or "fuse", to form a single heavier nucleus. This is usually accompanied by the release or absorption of large quantities of energy. Fusion is the process that powers active stars, the hydrogen bomb and experimental devices examining fusion power for electrical generation.
[81] Fuller, Steve; Kuhn, Thomas, "A Philosophical History for Our Times", University of Chicago Press, 2000
[82] In 1869, a Russian chemist Dmitri Mendeleev published the "Periodic Table". All known elements were arranged in order of their atomic weights and similar properties. The modern Table contains the atomic number (number of protons in the nucleus), element symbol, element name and atomic weight (approximate number of protons and neutrons in the nucleus).

synthetic methods that exist only for an instant, have a nucleus time rate differential that cannot survive in our time of one second per second. As universal time increases in the future, which it has done since the BB and must increase as the Universe expands; these elements will be stable in some future time zone. Nature has accomplished this same task of releasing energy in a seemingly less violent fashion by what we call the "weak force". Particles we term as radiation appear to be in sync with one second per second time passage and are released at certain half-life rates. These particles were trapped in the atoms nucleus eons ago, finally traversing through the atom's time zone into present day. An analogy would be a light photon taking millions of years to travel from the Sun's center to its surface before making the eight (8) minute trip to Earth. The length of time is different because the time zone at the Sun's center approaches zero as a limit and on Earth its one second per second.

As a black hole (BH) ingests an atom, the nucleus (held intact by the strong force) is squeezed into the singularity while the electron cloud (held together by a weaker force, the electromagnetic force), containing less than 0.001% of the atom's mass, is expelled in a tightly focused beam or jet along the axis of rotation of the BH's singularity. The size and velocity of these jets indicate the quantity of mass being ingested and the residual energy of the BH.

It is generally accepted that the gravity inside a BH is so strong that light photons, which have mass, cannot escape. There is at least one other logical reason that prevents light from escaping a BH: There are no particles intact that can produce light. Time flow near the BH's singularity approaches zero as a limit therefore one second is an eternity.

Time and gravity- The magnitude of gravity is dependent upon the location and density of the mass producing the gravity. Also, the gravitational field itself has mass and that mass, contributes to the total force exerted by the gravity field. Gravity has an inverse effect on the flow of time. As gravity increases time slows down. As

111

previously stated within a BH time approaches zero as a limit. There appears to be one exception to this rule. Gravity does not react with matter in the energy state, but only in the mass state. At least this appears to be true for the Strong Force.

Time is the single ingredient that transforms the four known fundamental natural forces into one unified force, gravity[83]. Here's how: within an atom the 'strong force' is present because the local time zone in the nucleus is near zero. If you apply Newton's gravitation equation[84] to the nucleus of an atom, replacing the 'gravitational constant' with a time of zero +, then gravity and the strong force appear to be the same. If the rate of time is increased then the weak and electromagnetic forces are in fact just simply different forms of gravity within different time zones. If time flow is increased to the present then gravity is as we experience it.

The energy stored in atoms, neutron stars, BHs and the Singularity, appears to decrease time within their respective time zones in proportional to their respective matter. Within the latter three, gravity appears to be the only natural force present. Take the BH for instance; the weak, strong and electromagnetic forces only exist, in time, up to the BH's event horizon[85]. At this point the decrease in time appears to convert all the natural forces into one force, gravity.

Somehow gravity and time are intertwined to the point where one is the manifestation of the other. Gravity and all its sub-forces binds the Universe whereas time is the medium in which energy can convert into mass.

---

[83] It's conceivable that we have been seduced by gravity and ultimately the strong force is the "unified force." It appears that the nuclei of atoms, BHs and the Singularity are intact because of the strong force; therefore gravity could be an aberration?

[84] $F=Gm1m2/d2$; G=rate of time flow instead of the gravitational constant; m1=nucleus (mass); m2=to electron (mass) and d2= distance between the nucleus and electron mass. Therefore the force of attraction (F) = the strong force which = gravity. $F= (0+) m1m2/ (0+) 2$.

[85] The event horizon is the threshold where nothing can escape.

The "current grand unification theories" (GUT) of forces include the two nuclear forces, the strong and weak, and the electromagnetic force but not gravity. The strong force is the force that binds together the particles in the atom's nuclei. The weak nuclear force is responsible for radiation. The electromagnetic force only acts between charged particles and is the transmitter of information between molecules/cells, etc. Electricity is a product of the electromagnetic force. While the gravitational force is the weakest of the four forces by many orders of magnitude, it interacts over huge distances between all particles.

With an increase in temperature and pressure the two nuclear forces weaken, with the weak force petering out first. With these conditions the electromagnetic force grows stronger.

Gravity has an inverse effect on the flow of time. If you use a BH as a model the two nuclear forces dissipate before the BH's horizon is reached, whereas the electromagnetic and gravity increase in magnitude. It's conceivable (see footnote 11) yet not proven that we have been seduced by gravity and ultimately the electromagnetic force (the force that transmits information) is the "UFT." Just maybe all information is not lost in a BH but recorded by the electromagnetic force?

Gravity and time appear to be more intertwined than time and the electromagnetic force. Therefore gravity could be the manifestation of time.

ENTROPY

"It was not easy for a person brought up in the ways of classical thermodynamics to come around to the idea that gain of entropy eventually is nothing more or less than loss of information". — Gilbert Newton Lewis

One reason that time has a direction is because there is a non-reversible process in the Universe called ***entropy***. As the Universe ages entropy increases. When the energy (matter) from the BB reaches equilibrium entropy and time will decrease instantly causing the "big crunch" that will produce a new Universe.

Astronomers[86] studying more than 200,000 galaxies have measured the energy generated within a large portion of space more precisely than ever before. They say that: "This represents the most comprehensive assessment of energy output of the nearby Universe. They confirm that the energy produced in a section of the Universe today is only about half what it was two billion years ago and find that this fading is occurring across all wavelengths from the ultraviolet to the far infrared. The Universe is slowly dying".

Entropy is a measure of disorder and is referred to as an "emergent property[87]" because it only emerges when a system becomes sufficiently complex. Since the Universe is a complex closed system, all natural processes are irreversible changes that can only continue because of time. In a closed system energy decreases as entropy increases with time. The natural processes that define matter appear to always conform to the laws of thermodynamics. These processes are irreversible, as far as we can surmise based on current knowledge.

Black Holes (BH) are the penultimate form of entropy compared to the ultimate form, the Singularity. Once any object is in the zone of influence of a BH, entropy increases until the object reaches the

---

[86] *European Southern Observatory-ESO, 10 August 2015*
[87] *Primack, J. R., Abrams, N. E., "The View from the Center of the Universe", Riverhead Books, 2006*

horizon or point of no return. This zone represents the maximum entropy. After the horizon point entropy begins to decrease rapidly as the object is reduced to its minimum particles that join the BH's singularity. Once inside the BH entropy is at its lowest. If the BH extracts radiation then entropy does not decrease to zero. The demise of the BH creates entropy which may rise as the BH evaporates.

Once the increase in time has converted all energy into mass, which has caused the Universe to "grow" to its ultimate size in order to contain the mass, entropy will have almost maximized. At this point in time energy in this closed system will be spread so thin, since moot of it hao boon oonvortod into mace, that the Univeree ehould cease expansion. My sense is that a state of equilibrium will exist for a few billion (current time base) years followed by a contraction phase which will occur at the speed of light relative to the existing time. Remembering that nearly all the maximum energy imaginable still resides in the Singularity, maximum entropy will occur when all energy is returned to same.

Another feasible scenario relates to the flow of time. Since time appears to be the catalyst, or by-product, of a conversion of energy into mass, what would happen if time simply stopped, became zero? Would all mass suddenly revert to energy? If so, the visible Universe would be gone in an instant!

Throughout each Universe the final repository of information is entropy. If there are cyclic Universes then entropy is also cyclic. Thus, the second law of thermodynamics only applies to each specific Universe from birth to death.

Back on Earth everything that was "alive" dies and, with time, breaks down to dust, or to the elements called atoms. Here entropy is at its maximum and, in most instances, begins anew as the elements are ingested into another living entity. This cycle has continued on Earth since life first appeared and continues to the present.

Everything in Nature seeks a state of lowest energy. That is why water flows downhill, we turn to dust after we die, and the weight of snow pack causes avalanches. This hypothesis is true for all matter throughout the Universe. Even the gravity force that appears to make this all happen will in time become zero as its creator, mass, disperses because of entropy.

UNIVERSE-A LOGARITHMIC PROGRESSION

"If the whole universe has no meaning, we should never have found out that it has no meaning: just as, if there were no light in the universe and therefore no creatures with eyes, we should never know it was dark. Dark would be without meaning." C.S. Lewis (1869-1963)

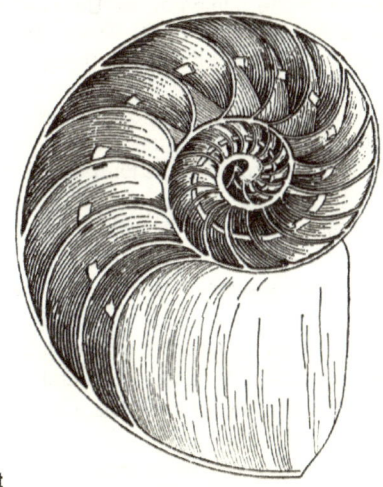

↑Past   Present →

My Universe

Each compartment represents the flow of time which causes the Universe to (face of the shell at any past time) expand to present day size. The "face" is always flat! The conceptual diagram records the past history of the Universe as time follows a logarithmic pattern that creates the future. The dimensions at the face are infinity in every direction.

From the sensing capability of Homo sapiens, developed within the same environment that is being sensed, the most logical model for the Universe appears to be the **"Spiral of Theodore of Cyrene"**. This natural form is called the Nautilus shell, which is a logarithmic spiral. If so, the spiral horn begins at the Singularity, the microcosm, which retains 99.99% of all energy, and spirals, infinitely outward, to create the macrocosm, our finite world. This design allow for a conservation of energy as the Universe expands. The face or cross section of the Nautilus is the visible Universe of today, thus the flat illusion. Scientists have looked back into the spiral for 13+ billion years, our clock but not the clock of the mass being observed. The farther back in time we observe the less mass we see, thus there is more invisible energy.

What we can agree on is the portion of the Universe that we can observe is less than a minute of the whole. There are hints that the observable Universe, the portion at the mouth (face) of the 'horn', is mass that has formed into spiral arms. We have observed what we call the "Great Wall" and other concentrations of mass to substantiate the spiral effect. In reality the Universe is so immense that we probably will never comprehend its form.

If time began at a near zero rate at the birth of the Universe and increased as the Universe expanded, its volume increase is 16 times faster than time measured (volume vs area). Yes the Universe is expanding but not at the rate that is observed but 16 times slower. The expansion rate appears to follow a logarithmic pattern, with infinite cycles, which appears to be a logical natural process. The logarithmic concept would account for the inflation period of growth.

Is the energy throughout the Universe an information source? Is the information about the Universe contained in matter that is in an energy state, which manifests into mass, with time, before entropy ultimately dominates as the current Universal cycle ends and another begins? The energy causes the Universe to expand as matter is created with the flow of time. Order/information is that the expansion/whole Universe looks the same from any direction before the future end game called entropy.

The Universal system, like everything else within its domain, is subject to the laws of thermodynamics. Once the whole energy system runs low on information, it cannot balance its account with matter, thus the Universe will fall apart (entropy) and only the renewal cycle driven by the Singularity can put it all together again.

Every corner of the Universe, from the microcosm to the macrocosm, has its own local time. From the flow of this respective time comes the mass state of matter that we can use to define the motion of any entity in relation to its neighborhood. What Einstein called the inertial frame in his special relativity theory (SR) could be the smallest locale with respective time zones that we can visualize until we have gained enough knowledge to delve deeper into the quantum world[88]. For now these frames exist only as a way to identify the passage of time.

The first known thought of a cyclic Universe can be traced to the Greek philosopher Plotinus, born in Egypt circa 205 CE. He suggested that an "eternal cosmic cycle" would produce infinitude of identical universes. Since Nature has never been the same in any scenario, it's conceivable that the 'eternal return' could produce an 'eternal cosmic cycle'.

Within nature everything begins anew, progresses through a cycle of change and eventually the whole ceases to exist, terminating into the original incremental building blocks/particles. Our Universe appears to be no different as, with the flow of time, it appears to be progressing from one methodical state to another as it ages, in an inescapable sort of way.

The Greek philosopher Thales (624 BC – c. 546 BC), who some believe to have begun the Greek philosophy movement, challenged the Greek Gods about who controlled the "natural" forces of the known world. Western civilization is built upon the knowledge and traditions of the ancient Greeks. They were the first to observe the world with questions about where did everything come from; where does everything go; why does everything work? Thales began the search for the laws of nature by separating the quest from the gods

---

[88] It appears that all particles, regardless of size, are made of smaller particles. This is true to the infinitum.

of mythology. These thoughts were the birth of scientific study about the world.

Pythagoreanism[89] followers believed that the Sun was the center of the Universe, but for more than 1700 years Aristotle's[90] views on the physical sciences profoundly shaped medieval scholarship. His influence extended well into the Renaissance, although it was ultimately replaced by Newtonian[91] physics.

---

[89] *Pythagoras of Samos (c. 570–c. 495 BC) was an Ionian Greek philosopher, mathematician, and founder of the religious movement called Pythagoreanism.*
[90] *Aristotle (384 BC – 322 BC) [1] was a Greek philosopher, a student of Plato and teacher of Alexander the Great.*
[91] *Sir Isaac Newton PRS (4 January 1643 – 31 March 1727 [OS: 25 December 1642 – 20 March 1727])[1] was an English physicist, mathematician, astronomer, natural philosopher, alchemist, and theologian.*

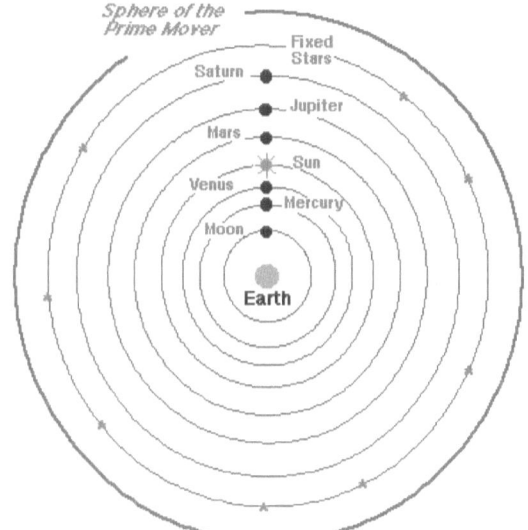

Aristotle's Universe

As the Universe ages, the rate of time increases as it's center of mass moves further away from the influence of the Singularity.   Since a very small amount of mass can store an incredible magnitude of energy, (remember Einstein's formula $E=MC^2$ where E- energy is equal to M=mass times the speed of light squared) within any given time zone, this has driven the growth/expansion of the Universe. Since energy is invisible to our sensing devices it could conceivably be mistaken for the elusive 'dark matter'. There appears to be residual energies throughout the Universe that will, in time, convert to mass as the Universe expands.

In the beginning, less than one percent of the energy contained in the Singularity was dispersed at near infinite velocity and near an infinite energy state to expand into the Universe. From Newton's point of view this event was caused by an irresistible force meeting an immovable object creating what is described as the Big Bang (BB). The collapsing of the last Universe was the irresistible force and the immovable object was the Singularity.

121

When all the energy ejected from the Singularity has converted into mass via time it can be said the Universe has reached equilibrium. However, at this point in the age of our Universe there is a continuing conversion of mass back to its original energy state within the nuclei (BH) of each galaxy. This process was started the moment the first galaxy formed. By the end of the Universe's life some BH will have continued to grow by ingesting mass and merging with other BHs.

The moment when the original energy, which was expelled from the Singularity, is exhausted, time will have no other function and the 'Big Crunch" will commence. The moment time approaches zero as a limit all mass will become energy again, and visual reality will no longer exist. There will be only one force, gravity. ... and another

EPILOGUE

"The questions that define physics are "how" and "why"; It is answers to the "why" questions that take us to limits of knowledge". Neville McMorris[92]

The lineage of today's science probably began to mature in the 16th century as a small gap began to open between science and philosophy. The apron strings of religion were beginning to loosen. This was a period that some historians believe began the awakening of present day science.

Kuhn[93] describes the extent of the "philosophical" dispute about the nature of science between the realist Max Planck (1858-1947) and the instrumentalist Ernst Mach (1838-1916). He considers the outcome of their dispute gradually to be taken for granted by philosophers of science, yet culminating in the Orwellian sensibility that he says is so integral to science's sense of its own history. (Fuller/Kuhn's book is worth a study of the philosophical difference between what Nature represents and realities that we have created.)

Throughout the history of science, there has not been a cordial relationship with religion. Until about the 19th century, the "Church" has been the constant authority on the free development and communication of scientific principles. However, over the centuries philosophy has not advanced in changing its doctrine to reflect the scientific discoveries that have been made. As a matter of fact, some prominent scientists, particularly physicists, believe that philosophy is near its intellectual end because of a lack of conforming to the knowledge that the world has gained about how Nature works. Martin Luther[94] is credited with the statement "that the biggest danger to religion was men who could reason".

---

[92] McMorris, Neville, "The Nature of Science", Fairleigh Dickinson Universe Press, 1989
[93] Kuhn, Thomas; Fuller, Steve, "A Philosophical History for Our Times", University of Chicago Press, 2000
[94] Luther, Martin (10 November 1483-18 February 1546) A German Priest, professor of theology and iconic figure of the Protestant Reformation.

A KNOWLEDGE ODYSSEY

A well documented conflict between religion and science involved Galileo Galilei (1564–1642), an astronomer, philosopher and physicist. He built the first telescope. After pointing it toward the stars, he soon discovered that Earth was not the center of the Solar System. The Sun was the center! Since this was against "Church" doctrine, he was put under house arrest for the remainder of his life. Had he not known the Pope nor had not been a famous scientists of that time, he most likely would have been burned at the stake.

By the second century B.C., the Greeks knew that the Sun was the center of the Universe. However, the church kept a lid on that knowledge for 1700 years.

Some philosophy teachers believe that the United States was created as a religious nation. There is a Treaty with Tripoli, dated 1796, that states that the U.S. was NOT founded as a religious nation. John Adams, an atheist, signed the treaty. Thomas Jefferson, who wrote the Declaration of Independence, was also an atheist; Benjamin Franklin was an agnostic, and George Washington never attended church. What these patriots believed in was that every individual had the freedom and liberty to believe in whatever they chose.

The motto "In God We Trust" was adopted in the mid 1950s, along with the phrase added to the Pledge of Allegiance; "under God". Today polls show that more people do not practice religion than do in the U.S., so maybe Stephen Hawkins (English scientist, physicist, and mathematician) is right when he writes in his new book, "The Grand Design", that philosophy has outlived its usefulness.

Today there are conflicts concerning what is taught in public secondary schools that involve the separation of church and state. This separation issue is very vague amongst most politicians and political parties and their philosophies. (e.g.-the religious right, etc)

I am not a student of philosophy, but I have studied the Greeks from about the sixth century B.C. They are the originator of the philosophy that our virtual world was built upon. This virtual world dominated society for about 18 centuries. Beginning with Galileo's questioning the science of the natural world, science has gradually exposed the curtain that hides the "Wizard of Oz", a.k.a. the Church.

Also, I do not wish this discussion of philosophy to become a referendum other than an honorable mention of its place in this world that we have created. Therefore I'll end these philosophical thoughts with a quote from Protagoras, 4th century B.C.: "Concerning the gods, I have no means of knowing whether they exist or not, nor of what form they are; for there are many obstacles to such knowledge, including the obscurity of the subject and the shortness of human life".

Scientists have also been a constraint to the advancement of pure scientific knowledge. Besides the claim for the success of "cold fusion" there are many questions about research results that are well documented[95]. A study reported by "ScienceDaily" (03/18/2015) has found some scientists are unknowingly and/or knowingly tweaking experiments and analysis of data to support conclusions that increase their chances of getting results that are easily published.

One of the best published works that "uncover the intellectual rivalries, petty jealousies, and faulty science behind one of the most famous experiment in the history of evolutionary biology" is Judith Hooper's "Of Moths and Men", Norton, 2002. The entire research was staged.

Another colossal example was the National Geographic Society's (NGS) efforts in trying to suppress the news from Kitty Hawk, NC, that the Wright brothers had successfully performed power flight,

[95]Hogan, James P., "Kicking the Sacred Cow", Simon & Schuester, 2004
Kruszelnicki, Karl, "Great Myths-Conceptions", A. McMeel, 2006
Smolin, Lee, "The Trouble with Physics", !st Mariner Books, 2006[95]

while the NGS's sponsored flight attempt with Langley in Virginia had continued to fail. There have been other reservations expressed concerning the NGS's sponsored expedition to the Poles.

In an article, "World Without Ice", published in the National Geographic Magazine, dated October 2011, the researchers present an analysis of the climate during the end of the Paleocene geologic period (56.1 million years ago) and into the Eocene period based upon analysis of core samples taken from deep in the Earth's crust. The conclusion of the researchers was that carbon[96] from an unknown carbon dioxide source caused this climate condition. This part of the conclusion is probably correct because the source of the carbon dioxide was from comets that visited Earth from the Jupiter region. However, this climate condition could NOT be applied to the argument that the entropy from burning fossil fuels, (e.g., coal fired power plants) in the 21st century, will, in time, produce a similar climatic result. This leap of science is unsustainable since the carbon dioxide in the upper reaches of the air column has always arrived from space and NOT from any activity on Earth's surface.

There are many more documented stories about questionable scientific results that have occurred throughout the history of science. One last "good read" is by James Hogan; "Kicking the Sacred Cow", Simon & Schuster, 2004

Science is getting close to the obsolescent limits based on current mathematical language. New languages must be discovered before we can begin to comprehend aspects of the quantum enigma in terms of understanding how to repeat its concepts. The current mathematical systems are not intricately flexible nor precise enough nor free of "constants" (fudge factors) to affect dependable results from similitude exercises that can be modeled with "acceptable" accuracy. More scientific energies must be committed to this effort

[96] NOTE: It appears that all of Earth's carbon is found in its crust, another clue that the element's source was from above.

than are currently being applied to creating more eloquent cosmological geometries! There is evidence that the laws of physics change as the Universe evolves[97].

One major example of the stagnation in science is the "superstring theory, model or hunch" that has not moved from square one in a quarter century. With no prospects beyond eloquences in sight, the subject still drives many scientific cultures.

In early November 2008, I was getting my annual eye exam with a doctor that I have known for some time. He was a military flight surgeon, and I flew military aircraft so we always had something to talk about during each visit. Our conversations continued as if no time had transpired in between. During this visit we were talking about the "dark" side of science which prompted him to show me three different vials of eye drops. "Research" has shown that each one is best for glaucoma. Each research endeavor was conducted by scientists that had PhDs in the field. The three efforts were financed by the respective manufacturer of the eye drops.

Science has had a dark side since its birth. The pressure on many scientists has biased the outcome of their work, either by constraints placed upon them by those who control the resources, such as time, funds and/or institutions. Too much research is conducted for the purpose of proving a preconceived notion or jerry-rigged to prove an outcome from a controlled experiment.

If a researcher's conclusions do not readily agree with the scientific community's accepted positions, it is almost never cited in scientist's papers. The" incestuous" scientific world tries to remain closed to those folk who dare to challenge their views. Fortunately, significant advancements have been brought about by pioneers who dare to challenge the "community". These are the scientists and engineers who chose to imagine outside the conventional scientific envelope!

---

[97] Brooks, Michael, "13 Things That Don't Make Sense", Doubleday, 2008

They did not collect data to help satisfy a "preconceived" scientific notion, but they allowed the analysis, of all the available data to lead them to a conclusion that has advanced the knowledge of science.

The acquisition of grant funds has become a cottage industry. The competition facing many grant requesters leads to the submission of grant applications containing fraudulent or questionable data. Sometimes by accident someone discovers useful new science that makes the process worthwhile. Most academic grants are used as a vehicle for students to receive advanced degrees, thus the payout will accumulate over the life of their careers. Most research can be classified as applied research to solve a known problem

I want to make one last point about nonexistent research results concerning DDT- the "bug' killer. As far as I know there was never any research that suggested that DDT was harmful to man and/or fowl. In Africa with the use of DDT the death rate from malaria and river blindness was nearly zeroed out. After special interest groups successful lobbying of the Federal Government to ban DDT, using fraudulent data (the eagles along the Potomac were having trouble with egg shells for more than a decade before DDT hit the market), the death rates in Africa had again climbed into the millions per year.

JUST 'CAUSE' SOMEONE SAYS IT DOESN'T MAKE IT SO! But it does add to your body of knowledge. In the early 1800's, a peasant asked an engineer: "How does a steam engine work?" The engineer drew several charts and plans to show the peasant how the fuel converted the water into steam and how the steam powered the steam engine. The peasant said: "I understand completely but where is the horse?"

Well I understand my passions completely, but I have no idea where the horse is!

Any unification of Universal structure will/must enter through the portals of time. The "Rosetta Stone" of the "unity theory[98]" must be the time key.

"If we end up with a coherent and consistent unified theory of the Universe, involving extremely complicated mathematics, do we believe that this represents "reality"? Do we believe that the laws of nature are laid down using the elaborate algebraic machinery that is now merging in string theory? Or is it possible that nature's laws are much deeper, simple yet subtle, and that the mathematical description we use is simply the best we can do with the tools we have? In other words, perhaps we have not yet found the right language or framework to see the ultimate simplicity of nature." Michael Atiyah (one of the greatest mathematicians of the second half of the 20th century.)[99]

The works of Nature are absolute. There are no constants, chaos, probabilities nor static states. The whole and all of its components must follow the same Natural laws. Time allows Nature to progress in a predetermined way throughout its components- microcosm and macrocosm. Natural selection through evolution has given Homo sapiens the ability to focus our mind's eye in whatever direction we freely choose. Our world[100] will, in time, die as the Sun dies, but nature will probably continue!

Everything that I write about means nothing to most of us because these theories do not affect our daily lives. We have absolutely no control over any aspect of Nature because we exist as part of it. Our quest will always be to gain knowledge of how Nature works. This quest began when the Greeks began to look into the "*scientica*" of how Nature works. They started questioning the "power" of their gods. As a result, the Greek gods slowly became part of their

---

[98] *Theory of everything!*
[99] *Woit, Peter, "Not Even Wrong", Basic Books, 2006*
[100] *The Solar System-see chapter on "ENTROPY".*

mythological history, thus, a non-entity in their daily lives. In time, "*scientica*" became 'science'.

Today the knowledge curve of science is advancing. As we discover Nature's secrets, we continue to add to the aggregate body of scientific knowledge that will greet our descendents.

ENGINEERING-MANAGEMENT GUIDELINES

These guidelines will provide:

1. Individual and corporate personnel the ability to make a characteristic diagnosis in relation to command, control, intra and inter-communications: Improve individual **negotiating** and **briefing** (oral presentations) strengths.
2. Provide a base for: General conceptual planning; Driving intra and overall organizational strategies; Developing courses-of-action; to plan, prepare, research and present an informational or decision briefing.
3. Improve individual self-esteem and organizational charisma.
4. Improve the functional aspects of individuals in any organizational entity with an objective/mission of creating one efficient "whole team" with one common unified goal.
5. Provide the wherewithal for individuals to analyze company objectives and related intra and inter-communications regardless of organizational levels of the individual.
6. Increase the individual's leadership traits and attributes and thus their corporate worth.
7. Provide the wherewithal for improving individual's skills in their ability to plan and participate as a negotiator for any objective or scenario from project solutions to management-union contract bargaining.
8. Provide the individual with the basic tools to make an analytical diagnosis of the command, control, communication and negotiation aspects of their respective organization from an individual and organizational team building perspective. This will result in providing excellence in their workforce.
9. Improving the work quality/quantity of their functional entity.
10. Improve their Espirit de Corps!
11. Establish accountability.
12. Provide a forum for individual and team creativity.
13. Provide a vehicle to increase the organizational and individual momentum.

MANAGEMENT ANALYSIS GUIDELINES

1. CORPORATE LEADER
- His or her goals; what does he/she expect?

2. DIVISION CHIEF
- What does he/she perceive to be the Corporate Leader's goals?
- What does he/she think the Corporate Leader expects from him/her?
- What does he/she expect from his/her subordinates?
- What does the Division Chief expect from him/herself?

3. FIRST LINE SUPERVISOR
- What does the First Line Supervisor perceive to be the Division Chief's goals?
- What does the First Line Supervisor perceive to be the Corporate Leader's goals? (Company direction as a whole)
- What does the First Line Supervisor expect from him/herself?
- What does the First Line Supervisor expect from his/her subordinates? Are your subordinates aware of your specific expectations?

Each level of management should develop a "what are you going to do for me (chain of command/management levels) during a period of time. e.g. - two weeks; 30 days; six months or one year; - whichever period is appropriate for the type of projects/mission of the organization.

NEGOTIATION (E.g. - Unions)
- Create goals/objectives: minimum accepted; maximum desired
- Create argument positions for each: pros and cons.

List goals in logical order for presentation using the following criteria: Intersperse "absolute" goals with "compromise" goals; Each absolute goal (one that will NOT be given up regardless) should follow a goal that will probably be offered up; A positive atmosphere of compromise and "everything's negotiable" must prevail even through several goals in reality are "nonnegotiable.

- Establish, create, and appoint a negotiator. One that can present the organization's positions but not someone that has the final decision responsibilities.
- Stage "mock" negotiating session(s) in house before "wet run" with opposite parties.
- During actual negotiating session maintain positive atmosphere of compromise and problem solving toward the common good. Each problem solution should be absolutely understood, in detail, by both parties, e.g.-written down in language understood by both parties.

A KNOWLEDGE ODYSSEY

www.ingramcontent.com/pod-product-compliance
Lightning Source LLC
Chambersburg PA
CBHW021953170526
45157CB00003B/973